Fernando Reinoso-Suárez,
Isabel de Andrés, Miguel Garzón

Functional Anatomy of the Sleep-Wakefulness Cycle: Wakefulness

With 44 figures

Springer

Fernando Reinoso-Suárez M.D., Ph.D.
Universidad Autónoma de
Madrid
Departamento de Anatomía,
Histología y Neurociencia
Arzobispo Morcillo 4
28029 Madrid
Spain
fernando.reinoso@uam.es

Isabel de Andrés
Universidad Autónoma de
Madrid
Departamento de Anatomía,
Histología y Neurociencia
Arzobispo Morcillo 4
28029 Madrid
Spain

Miguel Garzón
Universidad Autónoma de
Madrid
Departamento de Anatomía,
Histología y Neurociencia
Arzobispo Morcillo 4
28029 Madrid
Spain

ISSN 0301-5556
ISBN 978-3-642-14625-1 e-ISBN 978-3-642-14626-8
DOI 10.1007/978-3-642-14626-8
Springer Heidelberg Dordrecht London New York

Library of Congress Control Number: 2010938716

Cover design: WMXDesign GmbH, Heidelberg, Germany

Printed on acid-free paper

Springer is part of Springer Science+Business Media (www.springer.com)

Reviews and critical articles covering the entire field of normal anatomy (cytology, histology, cyto- and histochemistry, electron microscopy, macroscopy, experimental morphology and embryology and comparative anatomy) are published in Advancesin Anatomy, Embryology and Cell Biology. Papers dealing with anthropology and clinical morphology that aim to encourage cooperation between anatomy andrelated disciplines will also be accepted. Papers are normally commissioned. Original papers and communications may be submitted and will be considered for publication provided they meet the requirements of a review article and thus fit into the scope of "Advances". English language is preferred.
It is a fundamental condition that submitted manuscripts have not been and will not simultaneously be submitted or published elsewhere. With the acceptance of a manuscript for publication, the publisher acquires full and exclusive copyright for all languages and countries.
Twenty-five copies of each paper are supplied free of charge.

208 Advances in Anatomy, Embryology and Cell Biology

For further volumes:
http://www.Springer.com/series/102

Acknowledgment

This work was supported by Grants BFU2006-07430 and BFU2009-06991 from Spanish Ministry of Education and Science. We appreciate Ms. Carol F. Warren for revision of English language usage.

Abstract

Sleep is a necessary, diverse, periodic, and an active condition circadian and homeostatically regulated and precisely meshed with waking time into the sleep–wakefulness cycle (SWC). Photic retinal stimulation modulates the suprachiasmatic nucleus, which acts as the pacemaker for SWC rhythmicity. Both the light period and social cues adjust the internal clock, making the SWC a circadian, 24-h period in the adult human.

Bioelectrical and behavioral parameters characterize the different phases of the SWC. For a long time, lesions and electrical stimulation of brain structures, as well as connection studies, were the main methods used to decipher the foundations of the functional anatomy of the SWC. That is why the first section of this review presents these early historical studies to then discuss the current state of our knowledge based on our understanding of the functional anatomy of the structures underlying the SWC. Supported by this description, we then present a detailed review and update of the structures involved in the phase of wakefulness (W), including their morphological, functional, and chemical characteristics, as well as their anatomical connections.

The structures for W generation are known as the "ascending reticular activating system", and they keep and maintain the "thalamo-cerebral cortex unit" awake. This system originates from the neuronal groups located within the brainstem, hypothalamus, and basal forebrain, which use known neurotransmitters and whose neurons are more active during W than during the other SWC states. Thus, synergies among several of these neurotransmitters are necessary to generate the cortical and thalamic activation that is characteristic of the W state, with all the plastic qualities and nuances present in its different behavioral circumstances. Each one of the neurotransmitters exerts powerful influences on the information and cognitive processes as well as attentional, emotional, motivational, behavioral, and arousal states. The awake "thalamo-cerebral cortex unit" controls and adjusts the activation pattern through a top-down action on the subcortical cellular groups that are the origin of the "ascending reticular activating system".

List of Contents

Abbreviations

ACh	Acetylcholine
AChE	Acetylcholinesterase activity
CRF	Corticotropin releasing factor
CS	Central Raphe nucleus
C-type	Thalamic core type neurons
DAT-ir	DOPAMINE transporter immunoreactive
DR	Dorsal raphe nucleus
EEG	Electroencephalogram
EMG	Electromyogram
EOG	Electrooculogram
GABA	Gamma-aminobutíric ácid
Hcrt/Orx	Hypocretins/orexins
H1, H2, H3	Histamine receptors
MCH	Melanin-concentrating hormone
M-type	Thalamic matrix type neurons
LC	Locus coeruleus
LCα	Locus coeruleus alpha
LCC	Locus coeruleus complex
LdT	Laterodorsal tegmental nucleus
NPY	Neuropeptide Y
NREM	Non-REM sleep
Pα	Perilocus coeruleus alpha
Pb	Parabrachial nuclei
PpT	Pedunculopontine tegmental nucleus
PGO	Ponto-geniculo-occipital activity
RPO	Oral pontine reticular nucleus
SCoe	Locus subcoeruleus
SN	Susbstantia nigra
SWC	Sleep–wakefulness cycle
SWS	Slow wave sleep

rCBF	Regional blood flow
REM	Rapid eye movement sleep
vRPO	Ventral part or the oral pontine reticular nucleus
VTA	Ventral tegmental area
W	Wakefulness
ZI	Zona incerta
5-HT	Serotonin

The stages of the SWC alternate in a characteristic cyclic pattern throughout the night. The graphic representation of the SWC phases along a sleep recording is called a hypnogram, and it expresses the sequence and duration of the stages through the course of the night. The first REM sleep episode usually emerges 90–120 min after the individual goes to bed, and successive REM sleep episodes appear recurrently with an ultradian rhythm (4 or 5 more episodes) throughout the night. REM sleep episodes are shorter at the beginning of the night and become longer as the night progresses. This fact is correlated to the complexity and richness of the oniric contents; the longer and more complex dreams are those that are reported after waking from a REM episode in the late hours of the night. REM sleep is always preceded by an NREM episode; healthy individuals in normal conditions never shift directly from stage 0 (waking) to REM stage, and either a stage 3–4 episode or, more frequently, a stage 2 episode comes immediately before any REM sleep episode. Stage 3–4 episodes are more common at the beginning of the night, before the first REM sleep episode emerges. Afterwards, most of the NREM sleep is stage 2 episodes. Awakenings may happen from any sleep phase all through the night; generally they are short arousals that are not usually recalled the next day. Good sleep quality necessarily involves a scarcity of intrasleep arousals during the night and the nonprolonging of initial stage 0. If one is aroused frequently or if stage 0 is prolonged, insomnia may result.

The above features of the SWC stages describe the situation in a healthy young adult, but SWC organization is not stable over our life. Actually, age is the factor that is known to most affect SWC architecture. Total sleep time reduces progressively as we grow elder; thus, neonates spend about 17–18 h sleeping, the standard 7–8 h of sleep of the adult is already reached in teenagers, and from that age onwards sleep time declines gradually down to 5–6 h in old age. Newborns have high proportions of REM sleep, and delta EEG activity is not established until the third month after birth. Sleep spindles appear even later. Therefore, in the first months of life humans spend more time asleep than awake, and periodicity between sleep and waking is ultradian; however, the circadian regularity is soon acquired in the infant.

Stages 3–4 and REM stage diminish as we age, thus stage 3–4 is almost completely absent in old people and the REM sleep episodes are very short. In parallel, intrasleep arousals increase, and the circadian rhythm of SWC gradually becomes weaker so that older people sometime doze during the day time.

Most basic research in sleep has used cats. Besides the advantage of a quite stable brain size after 6–8 months of age, a great amount of behavioral and physiologic information has been gathered from these animals. Cats sleep abundantly with a marked polyphasic ultradian rhythm and their SWC has four stages, according to the polygraphic sleep scoring criteria defined by Ursin and Sterman (1981) in chronically implanted cats with cortical and subcortical electrodes for sleep recording.

Active W in cats shows a low-voltage, high-frequency EEG activity, sometimes referred to as desynchronized. Robust muscle activity is usually present in the EMG. Eye movements and frequent movement artifacts are present in the recording.

Relaxed W or Drowsiness (D) presents a 4–8 Hz intermediate- to high-voltage EEG activity over the posterior lateral cortex in at least half of the scoring epoch. The EMG recording shows a well-sustained activity and the EOG recording can show some eye movement.

NREM sleep (or slow wave sleep – SWS) shows sleep spindles (11–16 Hz; 0.5–2 s) over the sensorimotor cortex (3 per 30 s or more) and slows delta waves (0.5–4 Hz; $\geq$50 μV) in the posterior lateral cortex in at least 20% of the scoring epoch. Ponto-geniculo-occipital (PGO) waves may be observed in the lateral geniculate nucleus, especially in SWS periods immediately prior to REM sleep. Tonic activity of variable amplitude is present in the EMG. Some authors distinguish between SWS-1 (slow waves occupy less than 50% of the scoring epoch) and SWS-2 (slow waves occupy 50% or more or the scoring epoch). No EOG activity is usually observed, although some slow rolling eye movements can appear, especially during SWS-1.

REM sleep displays a low-voltage, mixed-frequency EEG activity. The EMG is at or near the noise level and shows no tonic activity, but there are occasional muscular twitches. REMs are present in the EOG recording and PGO waves are observed in the lateral geniculate nucleus recording.

Chapter 2
Revision of the Publications Describing the Anatomical Connections and Effects of Lesions and Electrical Stimulation of Brain Structures on the Sleep–Wakefulness Cycle

From the first bioelectric description by Berger (1929), which characterized W as an EEG of fast, low voltage waves (later called activated or desynchronized EEG) and sleep as an EEG of slow, high voltage waves (later called synchronized EEG), experimental researchers have looked for the brain structures responsible for the synchronization or activation of the EEG and conceived of them as sleeping or waking structures, respectively. The later discovery of REM sleep (Aserinsky and Kleitman 1953), also called paradoxical sleep, demonstrated that an activated EEG is also coherent with a sleep state; therefore, and this made it necessary to consider the characteristics of other bioelectrical parameters (EOG and EMG), in addition to EEG, to characterize the different phases of the SWC. For a long time, until the seventh decade of the past century, lesion and electrical stimulation of brain structures were the principal methods for exploring the foundations of the functional anatomy of the SWC. For this reason, we shall dedicate this chapter to these early studies and use them as the scaffold for describing the current knowledge on the functional anatomy of the different phases of the SWC (Fig. 2.1).

The replicable results obtained from well-controlled lesions or stimulation of brain structures in mammals have been always rewarding. Our experience is that the location and extension of specific lesions correspond quite closely with specific changes in the SWC, even in so complex a structure as the brainstem tegmentum. Figure 2.2 represents the changes in the SWC states in 8 sham-operated control cats and 18 experimental animals that had lesions located in different brainstem structures. Cats with similar lesions (thick vertical lines to the right of the figure) show similar variations in their SWC (Reinoso-Suárez and de Andrés 1976). On the other hand, in order to delineate the true functional transcendence of a lesion in the brainstem, it was very helpful to identify the different anatomical pathways that consecutively degenerated as a result of the lesion. Consequently, many functional sleep-lesion studies would include a parallel study of the connections of the lesioned area using anterograde and retrograde degeneration methods to examine the patterns of anterograde and retrograde degeneration in animals with a similar lesion.

Dealing with lesion and electrical stimulation experiments to study the functional anatomy of the SWC states, we should point out that no brain structure

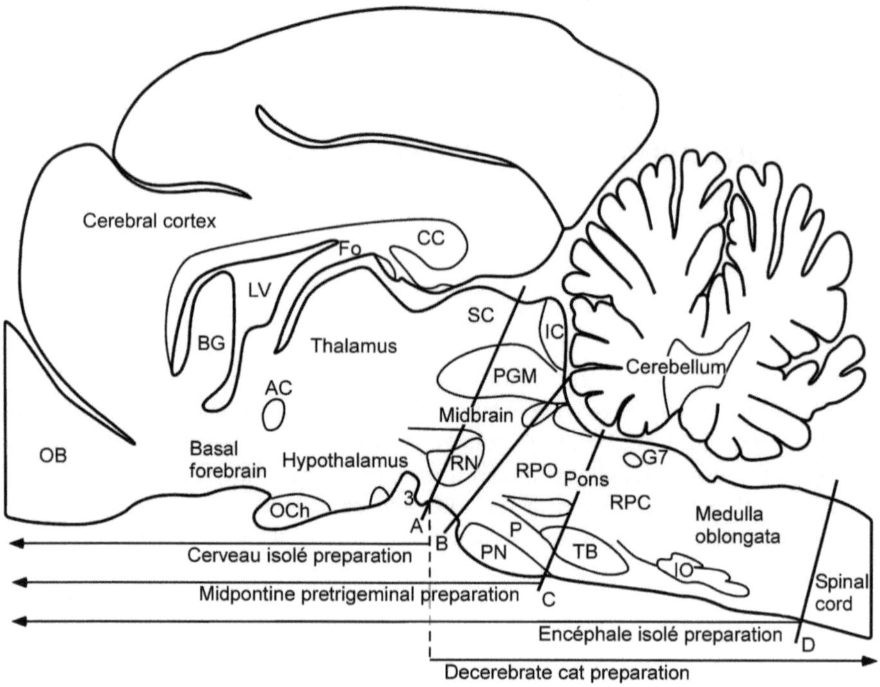

Fig. 2.1 Brain structures and experimental preparations in the cat. Schematic drawing of a parasagittal section of the cat brain illustrates the brain structures and experimental preparations resulting from different levels of brainstem transections discussed in this essay. (**A**) Intercollicular midbrain transection, (**B**) rostropontine pretrigeminal transection. The *cerveau isolé* and the *decerebrate cat* preparations are the rostral and caudal part, respectively, of these transections; (**C**) midpontine pretrigeminal transection: the *midpontine pretrigeminal* preparation is the rostral part, and the *midpontine cat* is the caudal part of this transection; (**D**) transection at C1: the *encéphale isolé* preparation is the rostral part and the *spinal cat* is caudal part of this transection. *3* third cranial nerve; *AC* anterior commissure; *BG* basal ganglia; *CC* corpus callosum; *Fo* fornix; *G7* genu facial nerve; *IC* inferior colliculus; *IO* inferior olivary nucleus; *LV* lateral ventricle; *OB* olphactory bulb; *OCh* optic chiasm; *P* pyramidal tract; *PGM* periaqueductal gray matter; *PN* pontine nuclei; *RN* red nucleus; *RPC* caudal pontine reticular nucleus; *RPO* oral pontine reticular nucleus; *SC* superior colliculus; *TB* trapezoid body

participates in only function and no function depends on a single brain structure (Reinoso-Suárez and De Andrés 1976). Each brain function is sustained by an extense, distributed, overlapping, and interactive neuronal network. In some brain functions, even the total organic unit can be involved. However, there may be a certain brain site or area that plays a major role in a specific brain function; this is a nodal and perhaps necessary link in the neuronal network supporting the function. The standard lesion and perhaps the stimulation studies may have attributed the responsibility to a given nucleus or area for a given

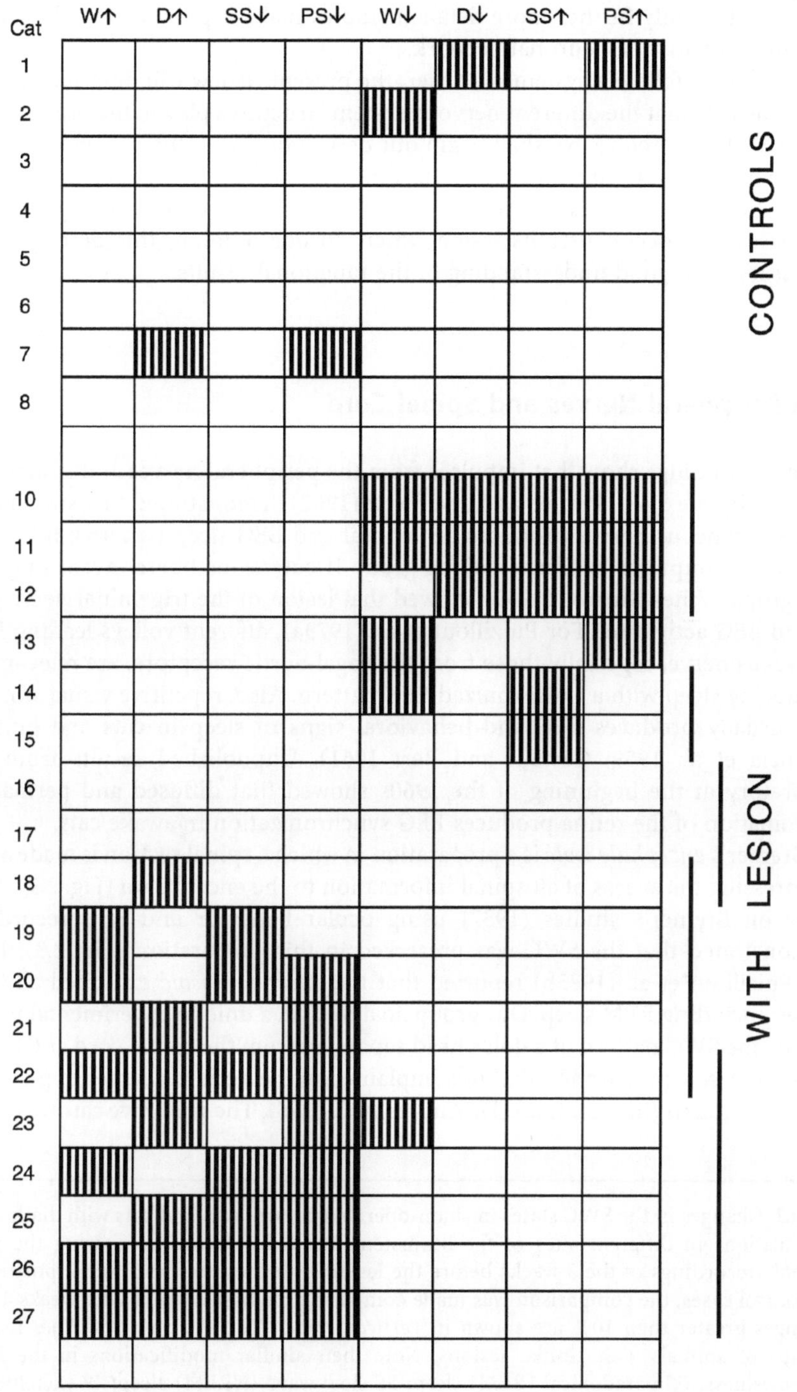
Cat
W↑ D↑ SS↓ PS↓ W↓ D↓ SS↑ PS↑
1
2
3
4
5
6
7
8
10
11
12
13
14
15
16
17
18
19
20
21
22
23
24
25
26
27
CONTROLS
WITH LESION

function, but only further more detailed studies make it possible to delineate all the components of a neuronal network.

In order to follow a systematic order, the present review will describe centripetally the role that the different nervous system structures play in the organization of the SWC. Therefore, we shall begin our description with the peripheral nerves and spinal cord and finish in the forebrain. Also, we will describe the anatomical connections of a precise area or structure and the subsequent degeneration processes that occur after its lesion, when, in our opinion, this description is necessary for a good understanding of the functional results.

2.1 The Peripheral Nerves and Spinal Cord

Different findings show that impulses from the peripheral nerves and spinal cord can modify the SWC. Pompeiano and Swett (1962) demonstrated that stimulation of the cutaneous nerves produces behavioral and EEG sleep patterns; this effect was due to impulses conducted along group II cutaneous fibers. Accordingly, in our group, Viñes-Morros (1959) showed that lesion of the trigeminal nerve gave rise to EEG activation. For Puizillout et al. (1973a), afferent volleys leading from the vagus nerve, especially those from the vagal-aortic receptors, are relevant for producing sleep with a synchronized EEG pattern. Also, repetitive visual stimulation usually produces EEG and behavioral signs of sleep in cats and humans (Mancia et al. 1959; Gastaut and Bert 1961). Unpublished results from our laboratory at the beginning of the 1960s showed that diffused and permanent illumination of the retina produces EEG synchronization in awake cats.

Bremer's *encéphale isolé* is a preparation in which a spinal section is made at C1, suppressing the access of all spinal information to the encephalon (Fig. 2.1). Very early on Bremer's studies (1937) using ocular behavior and EEG recordings demonstrated that the SWC was preserved in this preparation (Fig. 2.3). Later on, Puizillout et al. (1973b) reported that the *encéphale isolé* exhibited all SWC states, including REM sleep. Our group analyzed, in a unique experimental preparation, the SWC pattern of a dog's head separated from the spinal cord at C1 as an *encéphale isolé* preparation that was implanted in the neck of a larger dog with an intact neuroaxis (host animal) (De Andrés et al. 1976). The primitive carotids of the

Fig. 2.2 Changes in the SWC states in sham-operated control cats and cats with diathermocoagulations in different sites of the brainstem. Results obtained comparing the polygraphic recordings of the 3 weeks before the lesion with those of the 5 weeks postlesion. In control cases, the comparison was made comparing the first 3 weeks with weeks 4 to 8. Changes greater than 10% are shown in *vertical stripes*. The *thick vertical lines* include groups of animals with similar lesions. Note their similar modifications in the SWC. *D* drowsiness; *PS* paradoxical (REM) sleep; *SS* slow-wave (NREM) sleep; *W* wakefulness. Increase (↑), decrease (↓). From Reinoso-Suárez and De Andrés (1976)

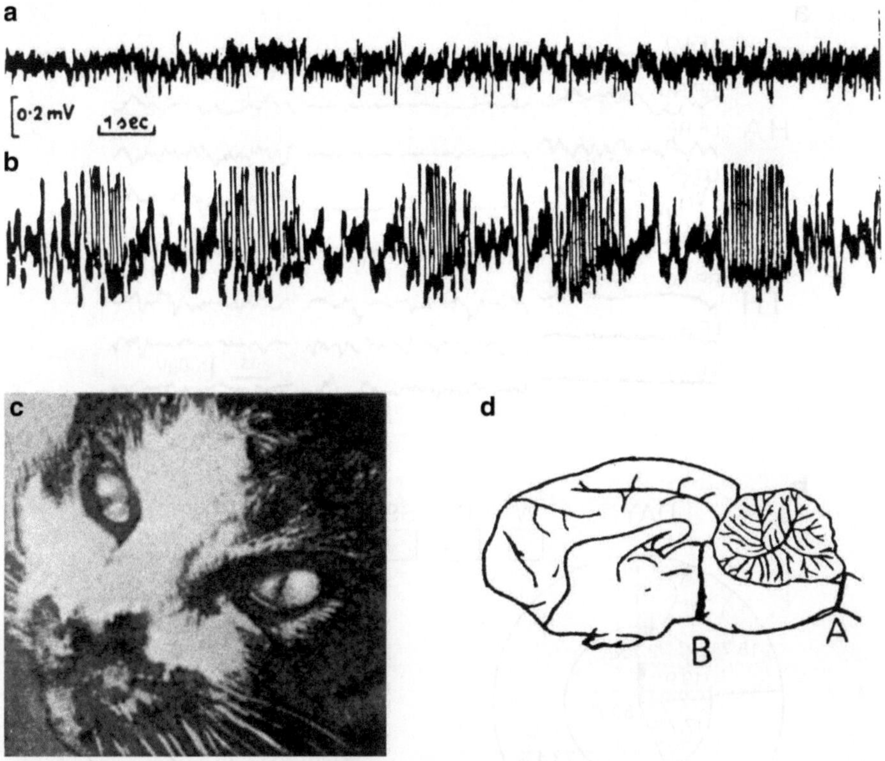

Fig. 2.3 EEG in the acute *encéphale isolé* and EEG and ocular behavior in the acute *cerveau isolé*. Following a spinal section at C1 (d_A), the *encéphale isolé* presents an alternation between EEG and ocular behavior of W (**a**) and sleep; however, following intercollicular section (d_B), the *cerveau isolé* presents a synchronized EEG pattern with high-voltage slow waves (**b**) and an ocular behavior (**c**) myosis that resemble those of intact cats during NREM sleep. From Bremer (1935)

implanted head were joined to the right internal and external carotids of the host; therefore, the implanted head received its blood supply from the host animal. The return circulation was achieved by joining the jugular veins of the implanted head to branches of the right jugular of the host. Both the implanted *encéphale isolé* and the host animal shared, therefore, a common circulation. Although the two brains were nourished by the same circulation, their SWCs were different in the temporal and intrinsic organization of their different sleep–wake states (Fig. 2.4a). In our dog *encéphale isolé*, all four phases (W, drowsiness, slow-wave sleep and REM sleep) were recorded over the course of the day and the night in a uniform sequence (Fig. 2.4b). The intact host animal had the dog's usual SWC with increased W during day and increased sleep at night. The medulla oblongata of the implanted *encéphale isolé* did not receive the impulses from the spinal cord and, in addition, it lacked those impulses from the vagus nerve including those from the vagal-aortic

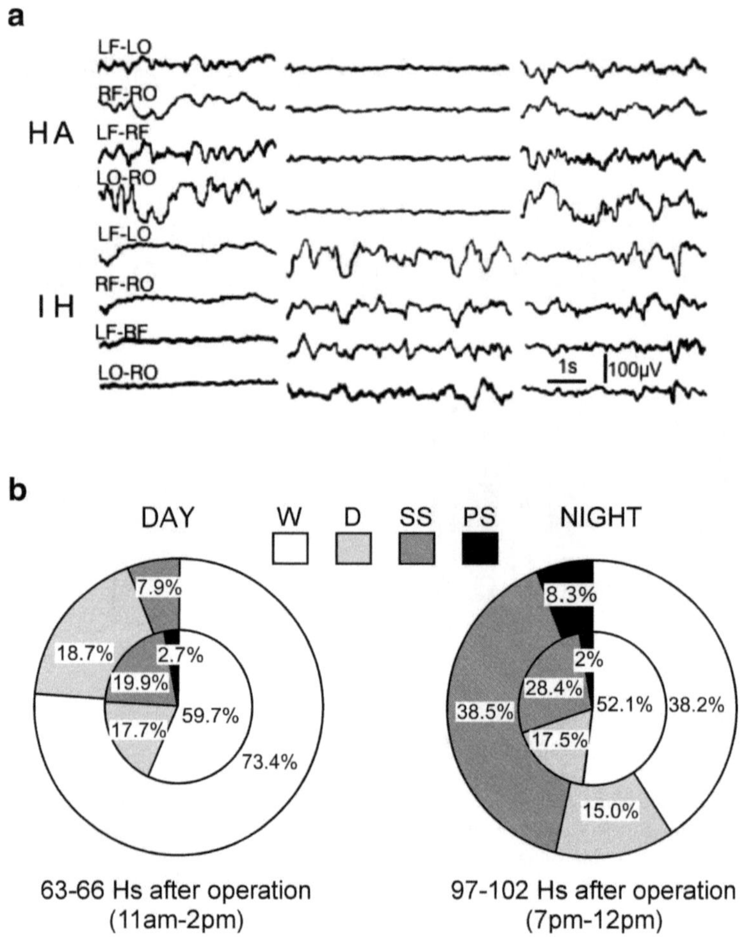

Fig. 2.4 Independence of the SWC behavior in two dog heads sharing a common circulatory system. (**a**) Various examples of the simultaneous EEG recordings from the host animal (HA) and the implanted *encéphale isolé* head (IH). Periods of independence and coincidence in their sleep wakefulness behavior can be observed. (**b**) Percentages of W (W), drowsiness (D), NREM sleep (SS), and REM sleep (PS) in the third day and fourth night after surgery. The larger outer field corresponds to the host animal and the small central fields to the implanted head: the host animal presented a nycthemeral rhythm with abundant W during day and NREM and REM sleep at night; meanwhile, the implanted *encéphale isolé* showed a monotonous SWC without difference between day and night. *LF* left frontal cortex; *LO* left occipital cortex; *RF* right frontal cortex; *RO* right occipital cortex. Modified (**a**) from Reinoso-Suárez and De Andrés (1976) and (**b**) from De Andrés et al. (1976)

and glossopharyngeal-carotid receptors that, as we have mentioned, are so relevant for producing sleep. The lack of these afferents might have produced the uniform and monotonous sleep–wake rhythm exhibited by the implanted dog *encéphale isolé*. Other findings can be mentioned here as examples of the role of spinal cord impulses related with the organization of sleep and W are: (1) the ascending synchronizing impulses that originated in the spinal cord discovered by Hodes (1964) since electrocortical desynchonization follows spinal block, and (2) impulses whose suppression in chronic spinal cord hemisections and lesions in the nuclei of the posterior column change the response in the mesencephalic reticular neurons to stimulation of the medullary reticular formation (Mancia et al. 1974a, b).

We studied the anterograde degeneration in rats with hemisections of the first cervical segments stained with the methods of Nauta (1957) and Fink and Heimer (1967) [Fig. 2.10a (C1), b]. Our findings regarding the spinal cord projections to the encephalon were rather similar to those of Nauta and Kuypers (1958) in the cat; however, we also found bilateral projections in brainstem and some differences in the prosencephalon. In any case, we described projections reaching further than the diencephalon (Reinoso-Suárez and De Andrés 1976). The projections in the diencephalon were basically located in the thalamus at both the ventral and intralaminar nuclei, and the fibers were incorporated in the ventral supraoptic decussation. The projections were bilateral to the medullary, pontine, and mesencephalic reticular formations. Therefore, afferents from the spinal cord may bilaterally influence all the reticular brainstem structures responsible for the SWC. Thus, it is evident that a lack of impulses from spinal cord, as occurs with those from the peripheral nerves, can modify the operation of the central structures that are more directly related with the regulation of the SWC.

2.2
Medullary and Caudal Pontine Tegmentum

Gottesmann concluded in a review at the end of the past century that, although the medulla oblongata is crucial to sleep–waking mechanisms, its contribution is not easy to demonstrate because medullary neurophysiologic properties are often concealed by those of the more cranially located structures in the brainstem (Gottesmann 1999). The medullary and midpontine cats, that is, the preparations caudal to the rostromedullary or midpontine transections, may show stillness immediately after the lesion with a brief period of primitive arousal (Fig. 2.1). From the third week post lesion onwards, behavioral activation shows marked periodicity (Siegel et al. 1986). Therefore, when the medulla and the caudal pons are separated from the more cranially located brainstem structures, there is a capacity for organized activity and rest periods, possibly resembling very elemental waking and sleep states but, in absence of influences from structures higher in the brainstem, the medulla and caudal pons are unable to generate true NREM or REM sleep signs because the full expression of these states cannot occur in chronic medullary and midpontine cats.

Moruzzi and Magoun (1949) demonstrated that electrical stimulation of the medullary reticular formation in lightly anesthetized cats gave rise to cortical EEG activation (Fig. 2.5). Also, chemical stimulation with adrenaline (Cordeau et al. 1963) or the cholinergic agonist carbachol in the most caudal part of the caudal pontine reticular nucleus and the medullary reticular formation increased behavioral and polygraphic patterns of waking in the unanaesthetized intact cat (Baghdoyan et al. 1984; Garzón 1996). Bonvallet and Dell (1965), by means of lesions at the medullary level that interrupted the projections from the area of the tractus solitarius nucleus, proved that impulses originated in this region could have an inhibitory action on EEG reticular activating structures. Magnes et al. (1961) and Berlucchi et al. (1964), using either stimulation or cooling of brainstem structures close to the floor of the fourth ventricle, at the level of the medulla oblongata and the caudal pons, also pointed to a hypnogenic role for this region. In our experience, unilateral dorsal lesions in the medulla oblongata, including the nucleus of the solitary tract, do not produce significant changes in the SWC (Reinoso-Suárez and De Andrés 1976); however, Reinoso-Barbero and De Andrés (1995) demonstrated that morphine and specific agonists of the Mu and Delta opioid receptors microinjected into the solitary tract nucleus enhanced all the polygraphic and behavioral manifestations of NREM sleep in a dose-dependent manner. All these results indicate the existence of structures in

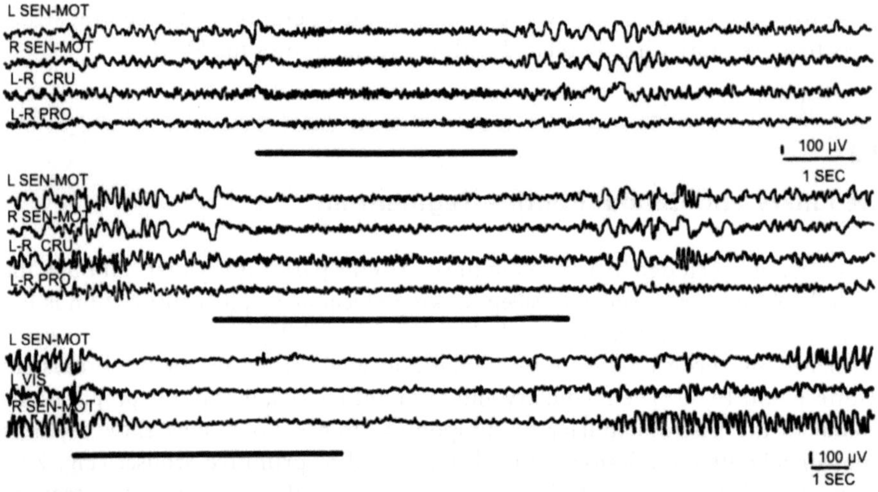

Fig. 2.5 Effects of medullary reticular stimulation upon EEG activity. The periods of reticular stimulation with electrodes situated in the *left* bulbar reticular formation are marked by a *heavy line* beneath the record. In every case, the stimulus produced cortical EEG activation. Upper and middle records: *encéphale isolé* cat with 7 mg/kg chloralose; stimulation with 1.5 V, 300/s. Lower records: intact cat with 50 mg/kg chloralose; stimulation with 3 V, 300/s. *L and R SEN MOT* left and right sensorimotor cortex; *L-R CRU* left to right cruciate gyrus; *L-R PRO* left to right proreus gyrus; *LVIS* left visual cortex. Modified from Moruzzi and Magoun (1949)

the medullary and very caudal pontine reticular formation that participate in waking mechanisms as well as the participation of the solitary tract nucleus in the induction of NREM sleep when an adequate stimulus is used. However, as well as the region of the solitary tract nucleus, there are many early experimental evidences indicating that the caudal pons and medulla have actions in NREM sleep. In our laboratory, Viñes-Morros (unpublished results, 1958) demonstrated that small lesions caused by diathermocoagulation in the caudal pontine tegmentum produce bilateral activation of the EEG. Later, Reinoso-Suárez et al. (1962) and Camacho-Evangelista and Reinoso-Suárez (1964) showed the existence of ascending EEG synchronizing influences arising from the nucleus reticularis pontis caudalis and neighboring areas, since small lesions at this level or in its ascendant projections produced continued sustained bilateral activation of the EEG (Figs. 2.6 and 2.7). These results were consistent with those of Batini et al. (1958, 1959) in the *midpontine pretrigeminal* preparation, which presented low-voltage EEG waves and ocular signs of alertness after a brainstem transection placed just in front of the trigeminal nerve roots and the caudal pontine reticular nucleus (Fig. 2.8). Also, Magni et al. (1959) reported a desynchronized EEG after pharmacological inactivation of the lower brainstem in the *encéphale isolé* cat. After small lesions in the caudal pontine tegmentum,

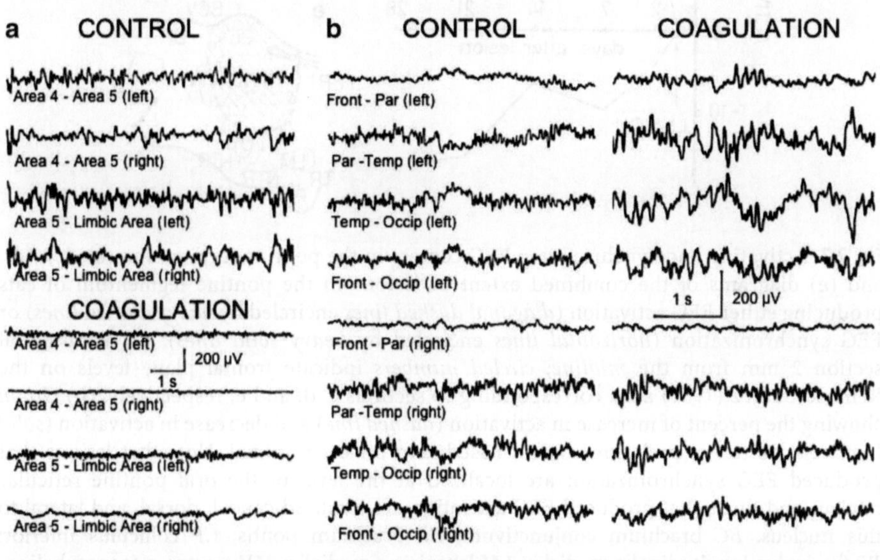

Fig. 2.6 EEG modifications after diathermocoagulation on the pontine tegmentum. (**a**) EEG recordings before (control) and after coagulation in the rostral half of the *left* caudal pontine reticular nucleus; note the bilateral activation of the EEG after the lesion. (**b**) EEG recordings before (control) and after the coagulation in the *left* oral pontine reticular nucleus; note the EEG synchronization, which is more conspicuous in the ipsilateral side after the lesion. Front (Frontal), Occip (Occipital), Par (Parietal), Temp (Temporal) cortices. Modified from Reinoso-Suárez et al. (1962)

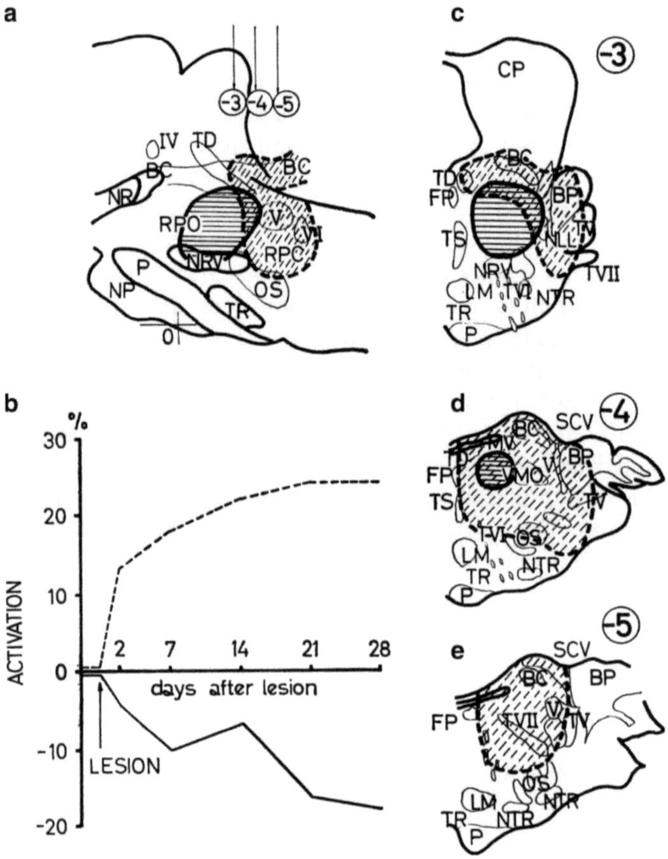

Fig. 2.7 Activating and synchronizing EEG centers in the pontine tegmentum. (**a**), (**c**), (**d**), and (**e**) diagrams of the combined extent of lesions in the pontine tegmentum of cats producing either EEG activation (*diagonal dashed lines* encircled by *heavy dashed lines*) or EEG synchronization (*horizontal lines* encircled by *heavy solid lines*). (**a**) Parasagittal section 2 mm from the *midline*; *circled numbers* indicate frontal plane levels on the Reinoso-Suárez (1961) atlas corresponding to sections c, d, and e, respectively. (**b**) *Graph* showing the percent of increase in activation (*dashed line*) and decrease in activation (*solid line*) relative to the prelesion control-based zero for each animal. Note that lesions that produced EEG synchronization are localized at the level of the oral pontine reticular nucleus and those that produced EEG activation are localized caudal, dorsal, and lateral to this nucleus. *BC* brachium conjunctivum; *BP* brachium pontis; *CP* colliculus inferior; *FP* fasciculus longitudinalis medialis; *LM* lemniscus medialis; *MV* tractus mesencephalicus nervi trigemini; *NLL* nucleus lemnisci lateralis; *NP* nuclei pontis; *NR* nucleus ruber; *NRV* nucleus reticularis ventralis; *NTR* nucleus corporis trapezoidei; *OS* oliva superior; *P* tractus pyramidalis; *RPC* nucleus reticularis pontis caudalis; *RPO* nucleus reticularis pontis oralis; *SCV* tractus spinocerebellaris anterior; *TD* nucleus tegmenti dorsalis; *TR* corpus trapezoideum; *TF* tractus tectospinalis. From Camacho-Evangelista and Reinoso-Suárez (1964)

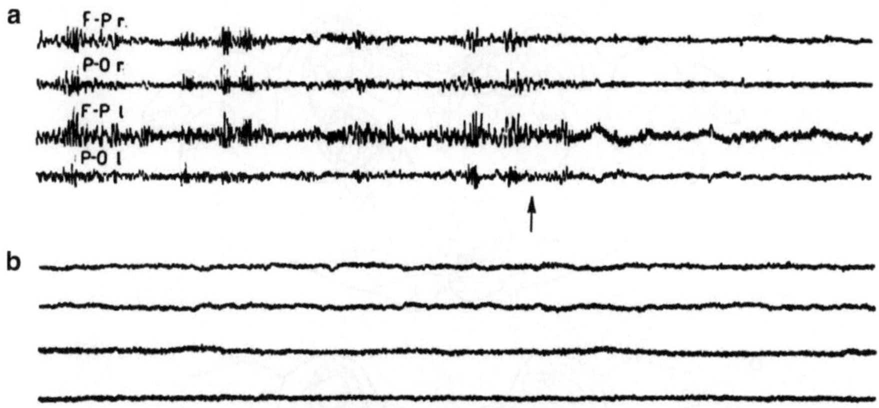

Fig. 2.8 EEG patterns in an anesthetized cat before and after a midpontine pretrigeminal transection. In a, the unrestrained intact cat situated in a quiet environment shows synchronized EEG with spindle bursts that are interrupted by a noise (*arrow*). In b, the same cat shows persistent EEG activation after a midpontine pretrigeminal transection. *F-P* fronto-parietal cortices; *l* left; *P-O* parieto-occipital cortices; *r* right. Modified from Batini et al. (1959)

we also found a decrease of NREM and REM sleep together with an increase in W and drowsiness (Zarranz and Reinoso-Suárez 1971; Zarranz 1972; Reinoso-Suárez and De Andrés 1976) (Fig. 2.9). Also, insomnia affecting both NREM and REM sleep occurred in humans after small vascular lesions of the pontine tegmentum (Forcadas and Zarranz 1994). Microinjections done in our laboratory with a small amount of a carbachol (mixed cholinergic agonist) solution in the rostral two-thirds of the caudal pontine reticular nucleus increased synchronized EEG activity (Garzón 1996; Reinoso-Suárez et al. 2001). Hemisection of the brainstem at the midpontine pretrigeminal level, done by Cordeau and Mancia (1959) and Rossi et al. (1963), demonstrated that no synchronized EEG recordings are present in the hemisphere on the side of the lesion at the beginning of a sleep period. All these observations strongly indicate that the structures of the medulla and caudal pontine reticular nucleus originate hypnogenic synchronizing impulses that reach the rostral brainstem and prosencephalon. These hypnogenic impulses, although bilateral, are more abundant in the ipsilateral side.

In addition, although the caudal pons and medulla cannot generate REM sleep in the absence of upper brainstem influences (Siegel et al. 1986), there are also experimental evidences indicating that the caudal pontine tegmentum and the medullary reticular formation participate in the organization of REM sleep, i.e.: (1) lesions in the caudal pontine tegmentum in unrestrained unanesthetized cats produce a statistically significant decrease in REM and NREM sleep (Zarranz and Reinoso-Suárez 1971; Zarranz 1972; Reinoso-Suárez and De Andrés 1976) (Fig. 2.9); (2) total transections of the brainstem at the caudal pontine level in cats and rats suppress behavioral and electrophysiological signs of REM sleep

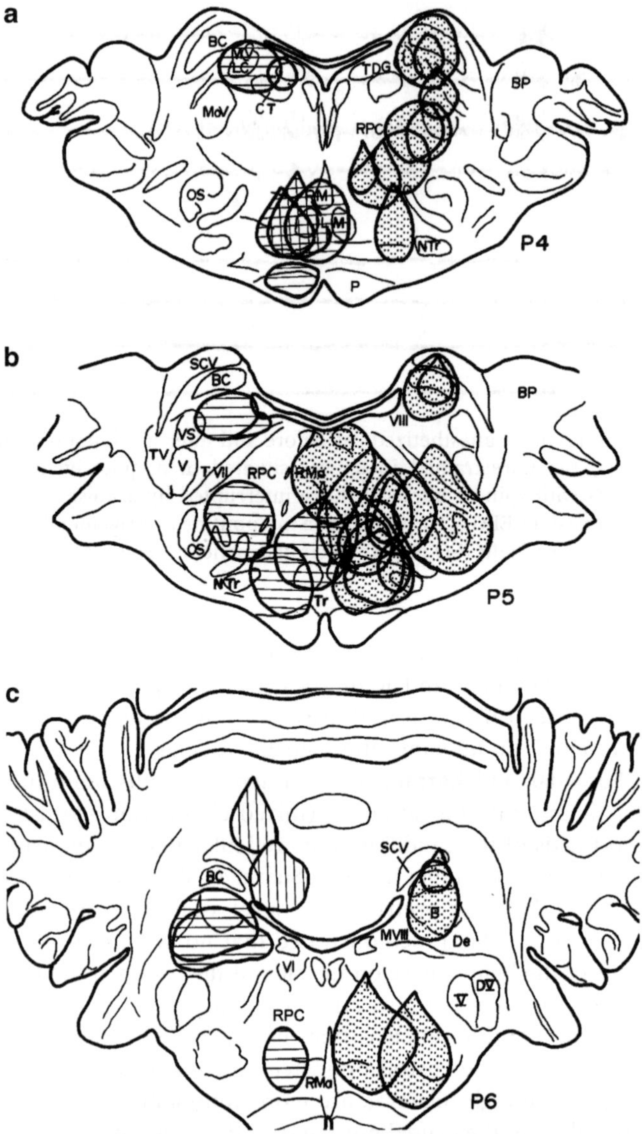

Fig. 2.9 Modifications of the SWC after limited diathermocoagulations in the caudal pontine tegmentum and the cerebellum. Coronal planes P4, P5, and P6 of the Reinoso-Suárez 1961 atlas with the location of restricted diathermocoagulation lesions in the caudal pontine tegmentum and the cerebellum that produce modifications of the SWC: *dotted areas* (*midline* and *right side*), statistically significant decrease of NREM and REM sleep; *vertically striped areas* (*left side*, caudal vermis in P6), significant increase of both sleep states; and *horizontally striped areas* (*midline* and *left side*), no significant changes in the SWC. Note that the sleep-decreasing lesions were mainly located in the caudal pontine

rostral to the transection (Gottesmann et al. 1995); (3) carbachol microinjections in the oral pontine tegmentum - which induce REM sleep in intact cats - are unable to induce this state after caudal pontine and premedullary transections (Vanni-Mercier et al. 1991); and (4) REM-on neurons have been described in the medullary medial reticular formation of the cat, where cytotoxic lesions produce a significant reduction in the amount of REM sleep and increased neck and limb muscle tone; it may be concluded that neurons of the medial medullary reticular formation of although they are not strictly necessary for REM sleep occurrence, they contribute to its generation; these neurons may have a particular importance for the regulation of muscle tone and inhibition of movement during this sleep state (Holmes and Jones 1994).

In our laboratory, serial hemisections made from caudal medulla oblongata to rostral midbrain in rats were processed with the silver impregnation method (Nauta 1957; Fink and Heimer 1967) to reveal the anterograde degeneration that was a consequence of the lesions (Fig. 2.10). The large lesions in these areas combined with the consecutive serial hemisections made it possible to delimit the projections from the zone under study as a whole; in addition, the step-by-step lesioning helped us to distinguish the degenerated fibers originating in caudal brainstem lesions from those originating in the lesioned area under functional study. Our anatomical results after unilateral lesions or hemisections in the caudal pons and medulla (Fig. 2.10, Pons Caud. and Medulla) which, we summarize next, revealed the morphological substrate for the physiological findings described above. Following lesions in the caudal pons and in the rostral medulla oblongata, it was possible to visualize and demonstrate bilateral projections to the pontine and midbrain reticular formation and the diencephalon since anterograde degeneration was found in these regions (Velayos 1971; Fairén 1973; Reinoso-Suárez et al. 1977) (Fig. 2.11a, b). Projections to parabrachial nuclei originate in the upper cervical cord and lateral medullary tegmentum. The caudal pons projects bilaterally, although most abundantly ipsilaterally, to the reticular structures situated rostrally, such as the locus coeruleus complex, the dorsal part of the substantia nigra, and the red nucleus. The projection to the zona incerta is most abundant on the contralateral side. Projections to the thalamus are abundant and occupy a large territory, terminating in the ventromedial, intralaminar, ventral, dorsomedial, and lateral thalamic nuclei. Degenerated formations were also observed in subthalamic and lateral hypothalamic structures extending rostrally to the lateral preoptic region and to the ventral supraoptic decussation. The mid-pontine

Fig. 2.9 (continued) tegmentum and in the brachium conjunctivum tract. *BC* brachium conjunctivum; *BP* brachium pontis; *LC* locus coeruleus; *LM* lemniscus medialis; *MoV* nucleus motorius nervi trigemini; *NTr* nucleus corporis trapezoidei; *OS* oliva superior; *P* tractus pyramidalis; *RMa* nucleus raphe magnus; *RPC* nucleus reticularis pontis caudalis; *SCV* tractus spinocerebellaris ventralis; *TD* nucleus tegmenti dorsalis; *Tr* corpus trapezoideum. Modified from Reinoso-Suárez and De Andrés (1976)

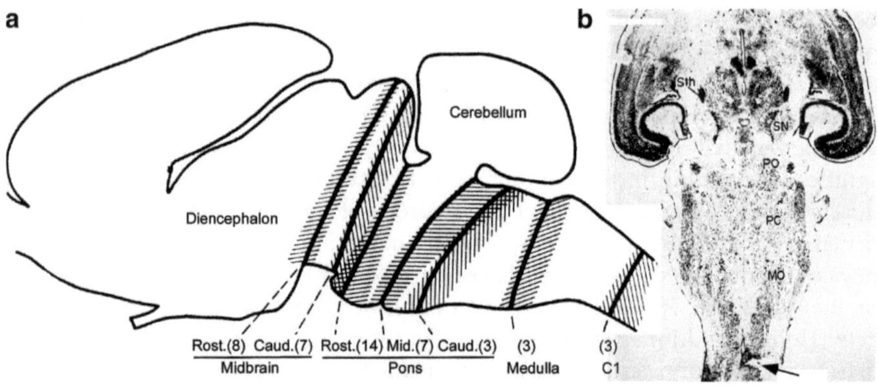

Fig. 2.10 Location of the rat brainstem transections discussed here. (**a**) The levels of the hemitransections are represented by *thick lines* on a drawing of the sagittal section of a rat brain: the total tissue destroyed after the different transections in all animals of each group is represented in the *striped areas*. The number (in *brackets*) of animals with a lesion at each level (midbrain, pons, medulla and spinal cord C1) is shown at the *bottom* of the figure. *C1* C1 segment of the spinal cord; *Caud.* caudal; *Mid.* middle; *Rost.* rostral. (**b**) Microphotograph of the horizontal section of a rat brain with a hemisection at the level of C1 (*arrow*). *MO* medulla oblongata; *PC* caudal pontine reticular nucleus; *PO* oral pontine reticular nucleus; *SN* substantia nigra; *Sth* subthalamic nucleus

hemisections produced very abundant degenerated ascending fibers, mainly ipsilateral ones, reaching the zona incerta and the lateral hypothalamic region; in this case, fibers in the dorsal supraoptic decusation could be followed to the contralateral lateral geniculate nucleus and superficial layers of the superior colliculus (Fig. 2.11c, d). The decussation of the degenerated fibers, after medullary and caudal pontine lesions, occurs at caudal pontine and rostral pontine levels. This gives rise to two observations: (1) terminal degeneration was located in the raphe nuclei and (2) the most rostral pontine hemisections produced fiber degeneration that was almost always ipsilateral as a consequence of the lesion to ascending fibers originated in the ipsi- and contralateral sides of caudal brainstem structures.

These anatomical findings from our laboratory support the physiological and retrograde degeneration data of Mancia (1969) who ascribed the decrease of sleep in the split brain pontine cat to the fact that this lesion suppresses the ascendant projections from medulla and pons that cross the rostral pons. They also support the finding of bilateral modifications in the EEG after unilateral lesions situated in the caudal brainstem, and that these modifications are more manifest on the side of the lesion. And, equally, they support the assertion by Moruzzi (1972) that the deactivating structures of the lower brainstem might act by both inhibiting the ascending reticular system or by counteracting its influence at the level of the diencephalon.

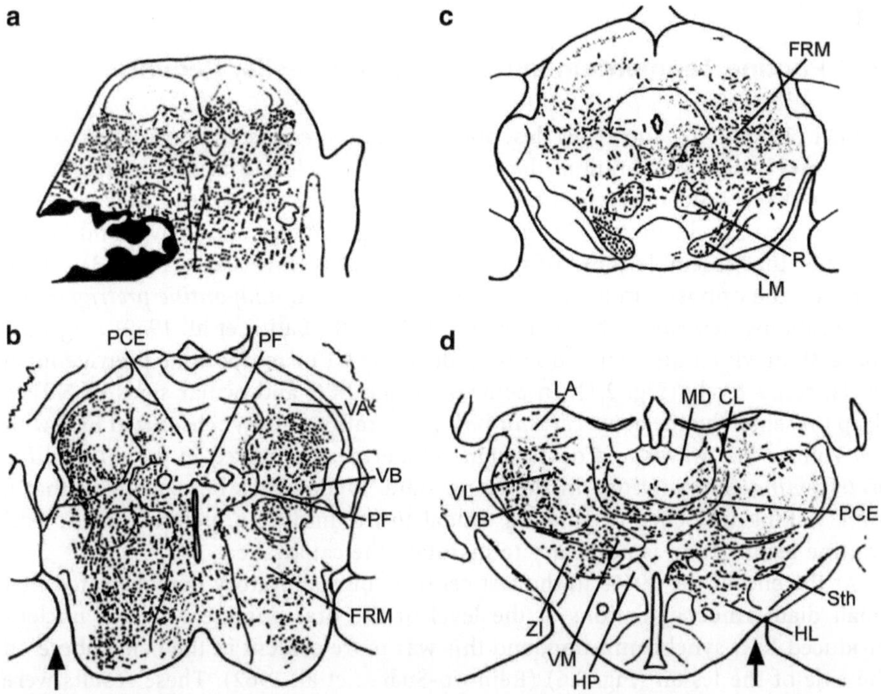

Fig. 2.11 Anterograde degeneration after hemitransections in the caudal pontine tegmentum. (**a**) and (**b**) Schematic drawings of horizontal sections of the brain of a rat with a caudal pontine hemisection on the *left side* (*arrow*) that affected the caudal portion of the caudal pontine reticular nucleus and the oral portion of the gigantocellular reticular nucleus. Degeneration spreads bilaterally in the brain stem and diencephalon. Terminal degeneration can be observed bilaterally, predominating on the side of the lesion, in the mesencephalic reticular formation (FRM), and thalamic nuclei parafascicularis (PF), ventralis anterior (VA), and paracentralis (PCE) as well as on the contralateral side in the ventrobasal complex (VB). (**c**) and (**d**) Schematic drawings of coronal sections of a rat brain with a middle pontine hemitransection on the *right side* (*arrow*) that damaged the rostral part of the caudal pontine reticular nucleus. A bilateral distribution of the degenerated fibers and terminal formations is to be observed in midbrain and diencephalon, ipsilaterally most abundant on the side of the lesion in the mesencephalic reticular formation (FRM), centrotegmental fasciculus, periaqueductal gray matter, deep layers of the superior colliculus (CS), lateral (HL) and posterior (HP) hypothalamic areas, subthalamic nucleus (Sth) and intralaminar (PCE, CL), lateral anterior (LA), and mediodorsal (MD) thalamic nuclei; and contralaterally most abundant in the contralateral side in the medial lemniscus (LM), red nucleus (R), superficial layers of the superior colliculus, zona incerta (ZI), ventral lateral (VL) thalamic nucleus, and ventrobasal complex (VB). Consistent bilateral degenerated terminals are found in the ventromedial thalamic nucleus (VM). Modified from Fairén (1973)

2.3
Oral Pontine Tegmentum and Superior Cerebellar Peduncle

Bremer (1935, 1937, and 1938) described the cat *cerveau isolé* preparation for the first time; in this preparation, the brainstem is transected at the intercollicular level or at the caudal mesencephalon (Fig. 2.1). The acute *cerveau isolé* preparation shows a synchronized EEG pattern with high-voltage slow waves and ocular behavior that resemble those of the cat during slow wave sleep (Fig. 2.3). These signs are the opposite in the acute *encéphale isolé* and *midpontine pretrigeminal* preparations, described above (Bremer 1937, 1938; Batini et al. 1959) (Figs. 2.3 and 2.8), in which the transection is made at the C1 or midpontine pretrigeminal levels, respectively (Fig. 2.1). In contrast to the EEG and ocular signs of NREM sleep presented by the acute *cerveau isolé* preparation, a desynchronized EEG with low voltage-fast waves and ocular signs of alertness occurred in the *midpontine pretrigeminal* preparations in the acute state. It may thus be concluded that a tonic waking influence originating caudal to the midbrain, perhaps in the oral pontine tegmentum, is necessary to maintain the cat awake.

At the end of the 1950s in the last century, our laboratory demonstrated that small diathermocoagulations at the level of the oral pontine reticular nucleus produced EEG synchronization, and this was more evident in the hemisphere on the side of the lesion (Fig. 2.6) (Reinoso-Suárez et al. 1962). These results were completed by Camacho-Evangelista and Reinoso-Suárez (1964), who demonstrated that small lesions in the oral pontine reticular nucleus and in the nucleus reticularis tegmenti pontis of Bechterew produced a bilateral increase in EEG synchronization that nevertheless was more noticeable on the side of the lesion. These results suggest a possible suppression of the waking influences that originate in these nuclei. In contrast, as mentioned before, lesions located immediately caudal to the previous lesions, at the level of the caudal pontine reticular nucleus, produced bilateral EEG activation, possibly by suppression of hypnogenic impulses originating in these nuclei (Fig. 2.7). All these findings indicate that the most caudal part of the ascending activating reticular system may be situated in the oral pontine tegmentum. The diatermocoagulation lesions at the level of the oral pontine reticular nucleus also produced a symmetrical bilateral increase of visually evoked-potentials in the visual cortices (Reinoso-Suárez 1962) (Fig. 2.12).

The oral pontine tegmentum is a complex area that requires a detailed analysis of each and every one of its structures: here the projections from caudal brainstem regions meet the superior cerebellar peduncle, which reaches the brainstem at this site; the structures located here are the oral pontine reticular nucleus, the rostral raphe nuclei (centralis superior and dorsalis), the locus coeruleus complex, and other important structures in the dorsal oral pontine tegmentum, such as the dorsal, the ventral and laterodorsal tegmental nuclei, and the pedunculopontine and parabrachial nuclei. To our understanding, this explains the complex manifestations observed after sections or

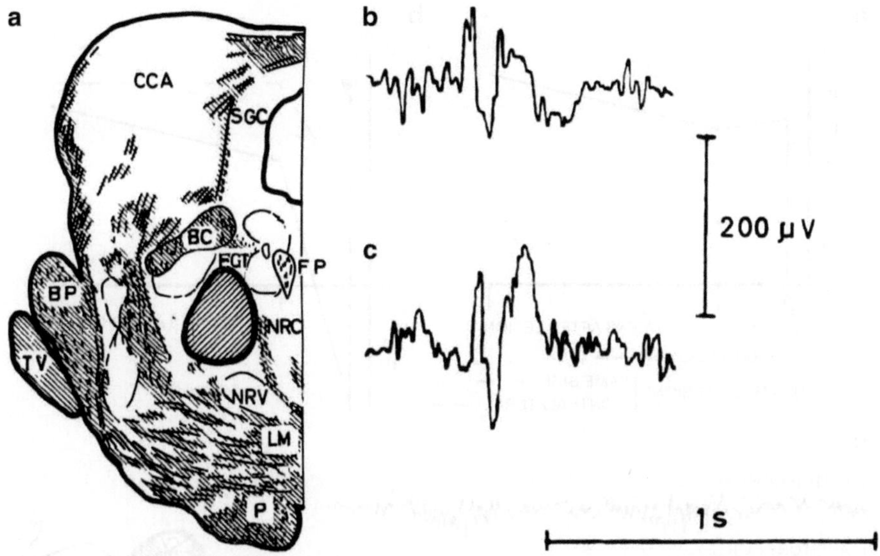

Fig. 2.12 The diathermocoagulation lesions at the level of the oral pontine reticular nucleus produced a symmetric bilateral increase of visual evoked potentials in the visual cortices. (**a**) Scheme of a lesion in the oral pontine reticular nucleus that produces both EEG synchronization and increased evoked potentials to visual stimuli in the visual cortex. (**b**) Visual evoked potential before pontine lesion and (**c**) visual evoked potential after oral pontine reticular nucleus diathermocoagulation represented in (**a**). *BC* brachium conjuntivum; *BP* brachium pontis; *CCA* inferior colliculus; *FCT* central tegmental fasciculus; *FP* medial longituninal fasciculus; *LM* medial lemniscus; *NRC* central superior raphe nucleus; *NRV* tegmental pontine nucleus; *P* pyramidal tract; *SGC* central gray matter; *TV* trigeminal nerve. Modified from Reinoso-Suárez (1962)

hemisections of the brain stem at this level (Cordeau and Mancia 1959; Rossi et al. 1963; Zernicki 1968). Our group reported that lesions in the long ascending tracts at the level of the oral pontine tegmentum produced EEG activation; bilateral activation of the EEG also followed uni- or bilateral lesions of the superior cerebellar peduncle that destroyed the brachium conjunctivum tract (Fig. 2.13a, b) (Camacho-Evangelista 1962; Camacho-Evangelista and Reinoso-Suárez 1964, 1965). This led us to infer that ascendant synchronizing impulses, originating in more caudal levels of the brainstem and from the deep cerebellar nuclei, were suppressed in these experiments. In the case of lesion of the brachium conjunctivun tract, the suppression of these synchronizing impulses may induce the generation of intermediate- to high-voltage theta band EEG activity, including the appearance of spindles in the thalamus, activities that can be recorded in the cortical areas of the projection from the thalamic nuclei where cerebellar projections end, principally in the contralateral motor and parietal cortices (Fig. 2.13c) (Camacho-Evangelista 1962; Camacho-Evangelista and Reinoso-Suárez 1965; Reinoso-Suárez 1992, 1993).

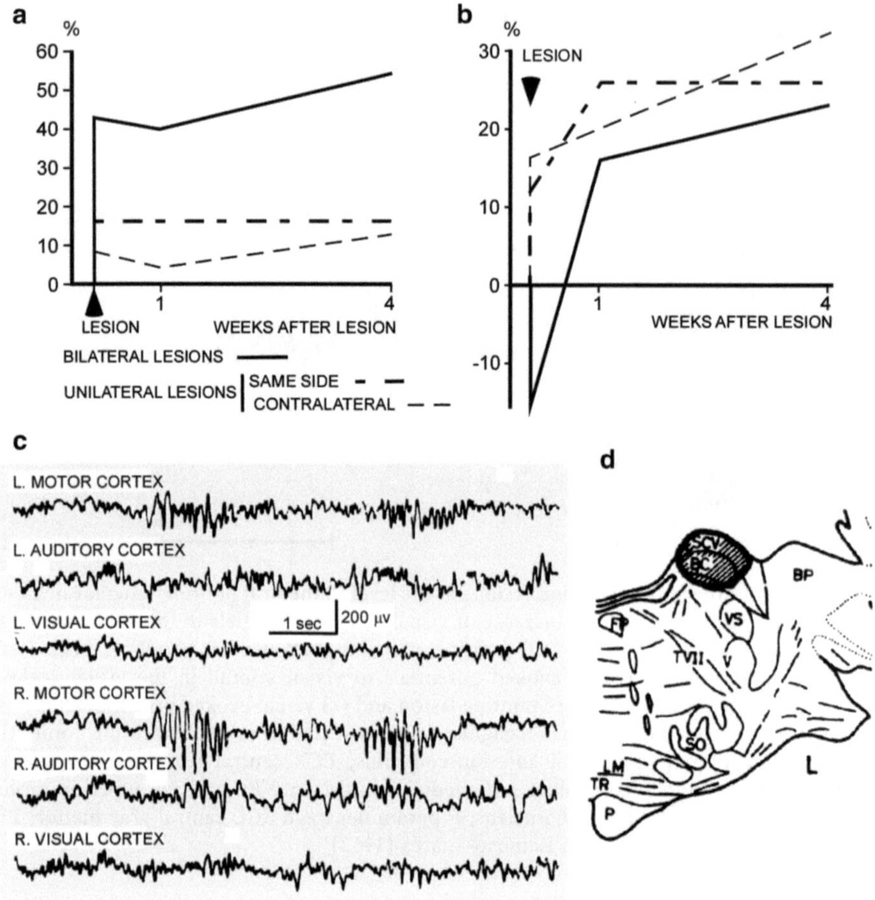

Fig. 2.13 Diatermocoagulation of the superior cerebellar peduncle produced EEG activation. (**a**) *Graph* indicating percentage increases in EEG activation after uni- and bilateral lesions of the brachium conjunctivum tract as shown in (**d**). Note the greater increase after bilateral lesions. (**b**) *Graph* showing the time-course of rapid frequencies in the motor cortex after the superior cerebellar peduncle lesions. An increase is produced immediately after inflicting a unilateral lesion and maintained during the 4 weeks that the experiment lasted. (**c**) Shows the appearance of synchronized groups of waves (7–13 c/s, bigger of 250 μV) in the *right* motor cortex after the lesion of the left brachium conjunctivum tract represented in (**d**). *BC* brachium conjuntivum; *BP* brachium pontis; *FP* medial longitudinal fasciculus; *LM* medial lemniscus; *P* pyramidal tract; *SCV* ventral spinocerebellar tract; *SO* superior olivary nucleus; *TR* trapezoid body; *TVII* facial nerve tract; *V* motor trigeminal nucleus; *VS* sensory trigeminal nucleus. Modified from Camacho-Evangelista and Reinoso-Suárez (1965)

Lesion to other structures of the oral pontine tegmentum produced complex and distinct results in the sleep–wakefulness states. Reinoso-Suárez (1971) described an increase in NREM sleep after small lesions in the rostromediodorsal part of the oral pontine reticular nucleus. If these lesions are larger, they also increase drowsiness and decrease W. Lesions in the tractus reticulorum tegmenti of Forel and the ascending reticular fibers surrounding the raphe nuclei, as well as the lesions in other long ascending tracts that pass through this level, increase W. Later studies showed that the lesions in the ventrolateral portion of the oral pontine reticular nucleus markedly decreased REM sleep (Gutiérrez-Rivas et al. 1978; De Andrés et al. 1975). Earlier lesion studies had attributed the organization of REM sleep to the structures located in the rostral half of the pons, specially the posterior third of the oral pontine reticular nucleus (see Zanchetti 1967 for a revision). In a more recent revision (De Andrés et al. 1989), we summarized the effects in the SWC obtained in our lab after mapping the oral pontine tegmentum with lesions located in the different structures of this region (Fig. 2.14). Unilateral diatermocoagulation in ventral and lateral areas of the oral pontine reticular nucleus (RPO) produced a specific and significant decrease of REM sleep (Gutiérrez-Rivas et al. 1978), while a significant increase of both SWS and REM sleep associated to a decrease in W followed unilateral lesions in the central part of the RPO (De Andrés et al. 1985). Unilateral diathermocoagulations in the dorsolateral pontine area, including the caudal part of the locus coeruleus complex, produced a specific and significant increase in REM sleep (Caballero and De Andrés 1986). These results indicated distinct actions by the RPO in SWC mechanisms: only the ventral-lateral part of this nucleus seemed to be specifically involved in REM sleep generation since the locus coeruleus complex seemed to exert an inhibitory influence on REM sleep. Other diatermocoagulation lesion studies performed in our laboratory to study the involvement of the oral pontine region in SWC mechanisms were focused on the raphe system located at this level. Lesions that destroyed between 13 and 100% of the central superior raphe nucleus produced an increase of W and a decrease of NREM and REM sleep; however, adjacent lesions in the paramedial region of the oral pontine reticular nucleus and passage fibers also decreased both NREM and REM sleep, although these decreases were associated to increased drowsiness; correlation coefficient analyses showed that the increase in W was the only state change that correlated significantly with the volume of central superior raphe nucleus destroyed; thus, this nucleus appears to be involved in arousal mechanisms rather than in direct sleep promotion (Arpa and De Andres 1993). The effect produced by the raphe nucleus lesion is compensated by the third or fourth week after the lesion. Finally, lesions of the lemnisci at the level of the oral pontine tegmentum only decreased sleep during the first week post lesion (Gutiérrez-Rivas et al. 1978).

We also performed lesions that destroyed the brachium conjunctivum tract in the superior cerebellar peduncle. Unilateral lesions of the brachium conjuncti-vum provoked retrograde degeneration in the central nuclei of the cerebellum (lateral and interpositus nuclei on the same side and certain areas of the

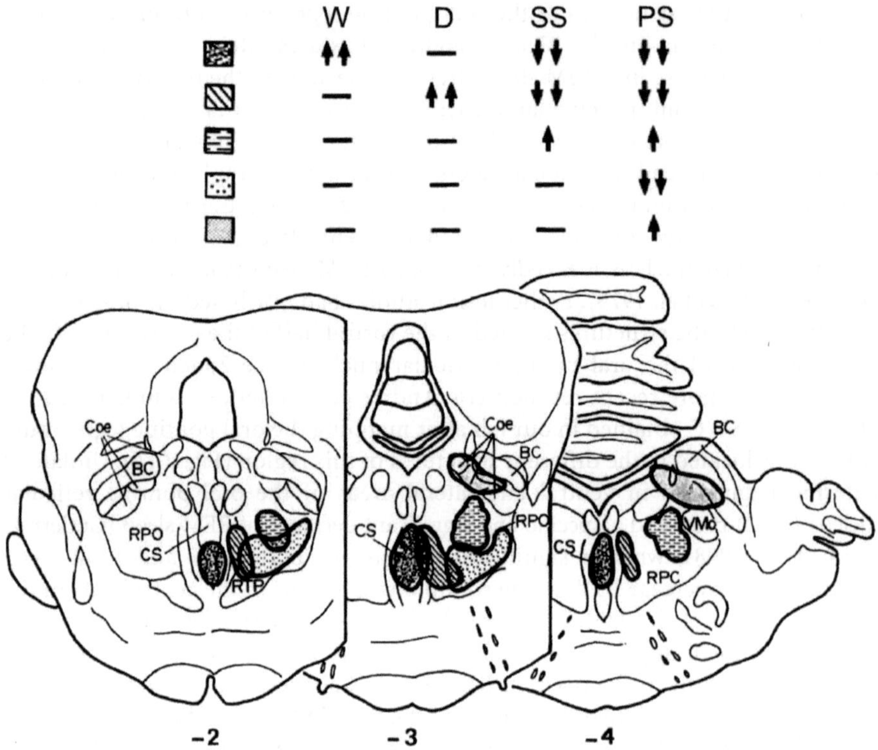

Fig. 2.14 Effects on the SWC phases produced by unilateral diathermocoagulation lesions located in different structures of the oral pontine tegmentum. At the top, the statistically significant increase or decrease of the SWC states produced by each group of lesions represented at the *bottom* with different textures on three brainstem coronal sections (-2, -3, and -4) of the Reinoso-Suárez (1961) stereotaxic cat atlas. *BC* brachium conjunctivum tract; *Coe* locus coeruleus complex; *CS* central superior raphe nucleus; *D* drowsiness; *RPC* caudal pontine reticular nucleus; *RPO* oral pontine reticular nucleus; *RTP* tegmental pontine nucleus; *SS* slow wave (NREM) sleep; *PS* paradoxical (REM) sleep; *VMo* motor nucleus of the trigeminal nerve; *W* wakefulness.↑, statistically significant increase; ↓, statistically significant decrease. Modified from De Andrés et al. (1989)

posterior region of the fastigius nucleus contralaterally); these lesions produced an increase in W and drowsiness together with a decrease in NREM and especially REM sleep that was expressed as a decrease in episode number and duration (Fig. 2.15) (De Andrés and Reinoso-Suárez 1979). In contrast, lesions of the cerebellar cortex and underlying white matter of the anterior cerebellar vermis increase REM and NREM sleep and decrease drowsiness and W (De Andrés and Reinoso-Suárez 1979). This effect is most conspicuous after large lesions of the posterior cerebellar vermis and the cortex of the cerebellar hemispheres, expressed by the very statistically significant increase in REM sleep, less pronounced increase in NREM sleep, and decrease in W, with a less marked

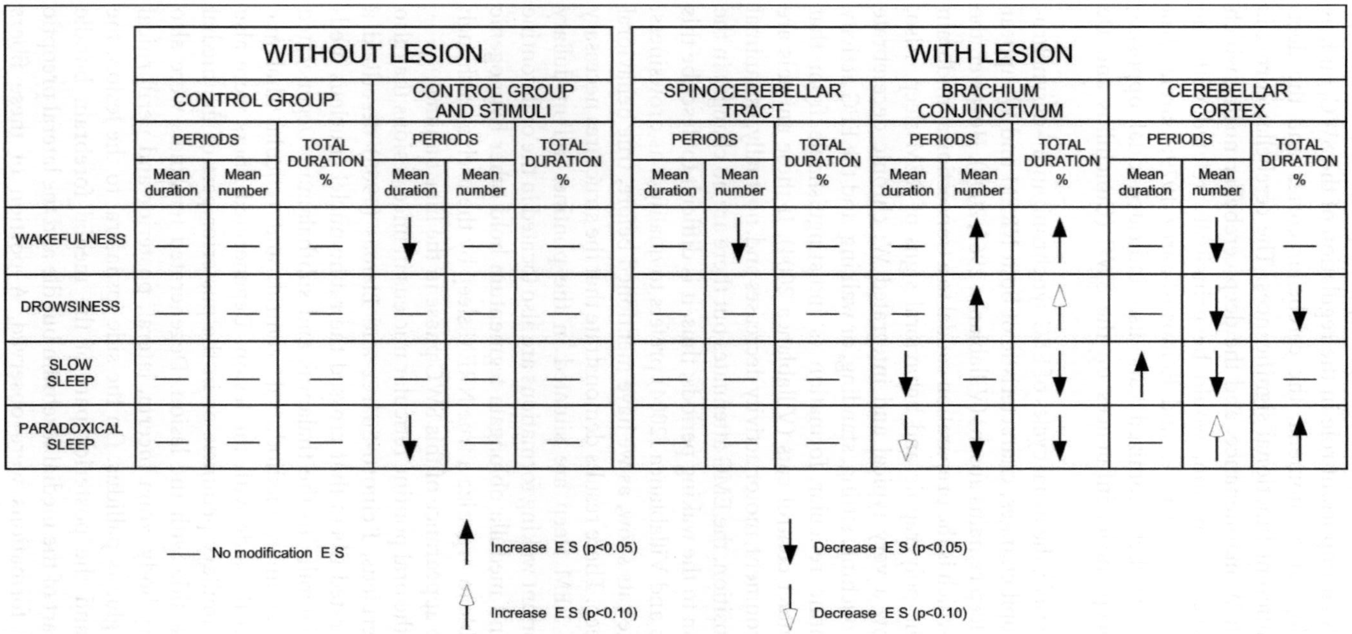

Fig. 2.15 Modifications in the mean duration and mean number of episodes per recording and total duration of W, drowsiness, NREM (slow sleep) and REM (paradoxical) sleep after different cerebellar lesions and in sham-operated controls. Note the great degree of stability in the SWC exhibited by the two control groups and by the group of animals with lesions in the dorsal part of the superior cerebellar peduncle (ventral spinocerebellar tract). The modifications in SWC parameters shown by cats with lesions in the brachium conjunctivum consisted in a significant decrease of NREM (Slow Wave) and REM (Paradoxical) sleep; these changes tended to be opposite to those shown by the cats with a lesion in the cerebellar cortex and white matter of the anterior vermis, the latter cats showing a significant decrease of drowsiness and a significant increase of REM Sleep. From De Andrés and Reinoso-Suárez (1979)

decrease in drowsiness (García Uría et al. 1980). It appears, from these findings, that the cerebellum plays an important role in the regulation of the SWC, and, as occurs in other physiological phenomena, the cerebellar cortex and the deep cerebellar nuclei have different functional significances. The cerebellar cortex is probably concerned with W maintenance, and the deep cerebellar nuclei, which probably have a hypnogenic function, would be principally related with the maintenance of REM sleep (De Andrés and Reinoso-Suárez 1979; García Uría et al. 1980). Studies in cerebellectomized cats also indicated dual opposing cerebellar cortex and deep nuclei influences on the SWC (Cunchillos and De Andrés 1982).

In the chronic *cerveau isolé*, the usual cycles of EEG synchronization–desynchronization together with pupil changes, characteristic of both NREM and W, appear progressively, but REM sleep remains absent (Villablanca 1965, 2004). However, the chronic decerebrate cat, which is the preparation caudal to a transection at midbrain level (Fig. 2.1), shows the polygraphic and behavioral signs of REM sleep. Also, decerebrate animals show a very typical and integrated W; chronic decerebrate animals may be found crouching, sitting, standing, or walking, and the EEG activity recorded from the pontine reticular formation is indistinguishable from that recorded during W in intact control cats (Villablanca 2004). If these animals are placed in a peaceful environment, motor activity decreases and, normally, the animal lies down in a random position, the EMG attenuates but there are no changes in the brainstem EEG in relation to the waking periods; thus, it is difficult to describe this state as true NREM sleep and Villablanca (2004) prefers to qualify it as drowsiness. However, the decerebrate cats show, as we have mentioned before, true behavioral and polygraphic REM sleep. These results demonstrate that the structures necessary for the organization of REM sleep are situated in the pontine and medullary brainstem and that important waking formations are also located in the oral pontine tegmentum. The pons and medulla oblongata tegmentum hold other hypnogenic structures that are not able to organize a true NREM sleep by themselves, although they do contribute to the appearance of this SWC phase in the intact animal.

After large lesions of the oral pontine reticular nucleus in mice, lesions that also destroyed the locus coeruleus, Reinoso-Suárez and Llamas (1968) described a great number of degenerated fibers that crossed the raphe nuclei (nucleus medianus) and ascended bilaterally to the thalamus and subthalamus. Degenerated formations in the intralaminar, mediodorsal, ventral, and reticular thalamic nuclei were observed on the side with the lesion. Degenerated fibers were also found in either the mammillary peduncle or in the posterior part of the medial forebrain bundle on the side with the lesion. Degenerated terminals were also noted in the mammillary body, zona incerta, lateral, posterior and perifornical hypothalamus, and the globus pallidus. On the side ipsilateral to the lesion, the mammillary peduncle and the posterior part of the medial forebrain bundle reunite in the anterior part of the medial forebrain bundle and the lateral preoptic region, where terminal formations were observed. A portion of these fibers traveled toward the diagonal band of Broca and the septal nuclei (Figs. 2.16).

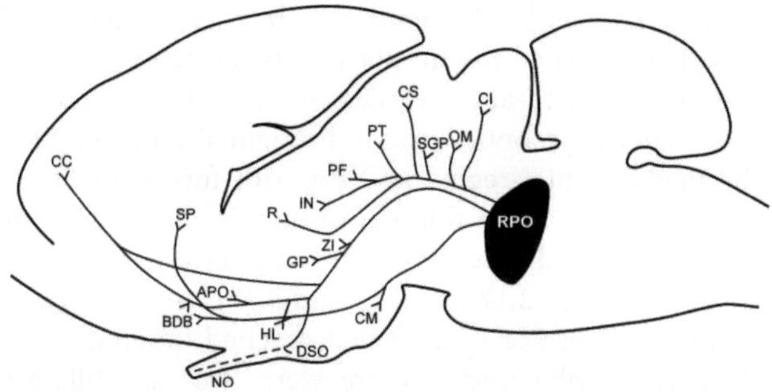

Fig. 2.16 Pattern of degenerated fibers after large lesions in the oral pontine tegmentum in mice. This sketch of a sagittal section of a mouse brain shows the ascending reticular fibers after a lesion of the oral pontine reticular nucleus (RPO) that also included the brachium conjunctivum tract, the locus coeruleus and the sensory trigeminal nucleus. A large number of degenerated fibers cross the oral pontine raphe with the consequent bilaterality of the projections to the midbrain tegmentum and diencephalon. Preterminal degenerated fibers are observed in the midbrain reticular formation but also in the inferior colliculus (CI), oculomotor nuclei (OM), periaqueductal gray matter (SGP), superior colliculus (CS), pretectal region (PT), thalamic nuclei parafascicular-centromedian group (PF), and intralaminar and midline (IN). Other fibers that originate in the oral pontine tegmentum project to the zona incerta (ZI), globus pallidus (GP), lateral hypothalamic area (HL), and lateral preoptic area (APO). Some fibers from this bundle pass through the dorsal supraoptic decussation (DSO) and reach the lateral geniculate nucleus, the pretectal region, and the contralateral superior colliculus. Others reach the cerebral cortex by means of the internal and external capsules. Finally, there are fibers that, following the path of the mamillary peduncle, incorporate to the medial forebrain bundle and advance to reach the preoptic region, the diagonal band of Broca (BDB), and the septal nuclei (SP). Modified from Reinoso-Suárez and Llamas (1968)

Other fibers advanced along the sublenticular or intralenticular portions of the internal capsule and external capsule terminating in the cerebral cortex. This was the first description in the literature of projections from the brainstem tegmentum to the cerebral cortex (Reinoso-Suárez and Llamas 1968). On the contralateral side, abundant terminal degeneration was observed in the oral pontine reticular nucleus and degenerated fibers in the medial lemniscus and brachium conjunctivun tract, possibly as respective consequences of the lesion of the principal sensory nucleus of the trigeminal nerve and the superior cerebellar peduncle; consistently degenerated terminals were found on the contralateral side in the ventrobasal and ventral medial and ventral lateral thalamic nuclei. From the medial forebrain bundle, on the side of the lesion, some fibers reached the dorsal supraoptic decussation and terminated in the contralateral pretectal region and lateral geniculate nucleus.

In our rostropontine hemisections in the rat (Fig. 2.10, Pons Rost.), which did not reach the midline, we confirmed the fiber degeneration in the thalamus, subthalamus, hypothalamus, preoptic region, cerebral cortex, and hippocampus

(Reinoso-Suárez and Llamas 1975; Reinoso-Suárez 1977; Reinoso-Suárez et al. 1974, 1975, 1977). The fibers to the hippocampus, most probably originating in the locus coeruleus (Pasquier and Reinoso-Suárez 1978), follow two different pathways from the lateral preoptic area and diagonal band of Broca: (1) some pass through the medial septal region and superior fornix and terminate in the dorsal hippocampus and (2) other fibers turn dorsorostrally around the genu of the corpus callosum to reach the cingulum and terminate in the dorsal and posterior hippocampus [Figs. 2.17 (label 1) and 2.18]. In the cases of rostropontine hemisections (Fig. 2.10, Pons Rost.) that reached the midline and damaged the dorsal and central raphe nuclei, there were two new bilateral groups of

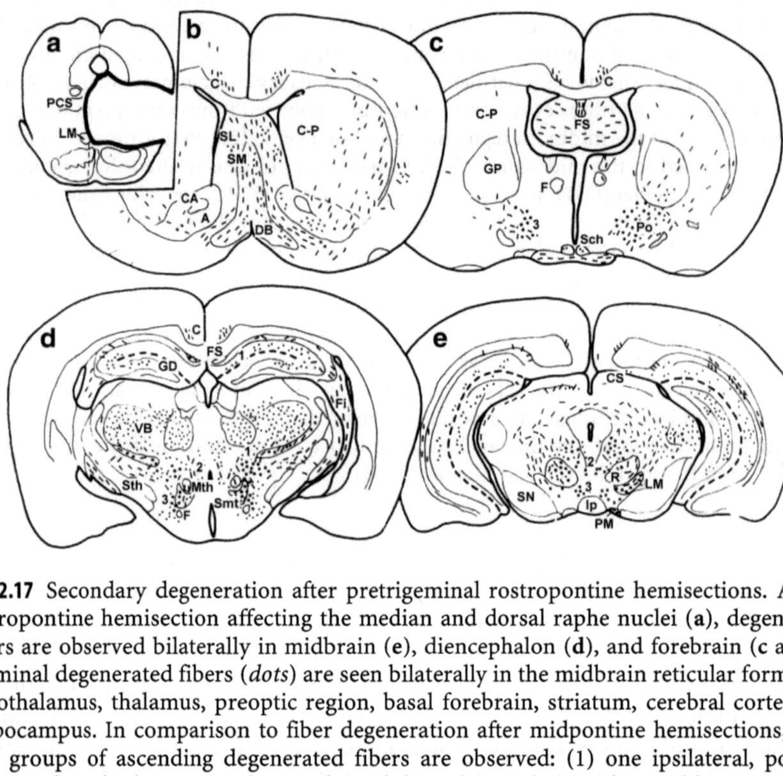

Fig. 2.17 Secondary degeneration after pretrigeminal rostropontine hemisections. After a rostropontine hemisection affecting the median and dorsal raphe nuclei (**a**), degenerated fibers are observed bilaterally in midbrain (**e**), diencephalon (**d**), and forebrain (**c** and **b**). Terminal degenerated fibers (*dots*) are seen bilaterally in the midbrain reticular formation, hypothalamus, thalamus, preoptic region, basal forebrain, striatum, cerebral cortex, and hippocampus. In comparison to fiber degeneration after midpontine hemisections, three new groups of ascending degenerated fibers are observed: (1) one ipsilateral, possibly originated in the locus coeruleus and two bilateral (2 and 3), with possible origin in the raphe nuclei. *A* accumbens nucleus; *C* cingulate fasciculus; *C-P* caudatus-putamen; *CA* anterior commissure; *CS* colliculus superior; *DB* diagonal band (Broca); *F* fornix; *FS* fornix superior; *GD* dentate gyrus; *GP* globus pallidus; *Ip* interpeduncular nucleus; *LM* medial lemniscus; *Mth* mamillothalamic tract; *PCS* superior cerebellar peduncle; *PM* mamillary peduncle; *Po* preoptic region; *R* red nucleus; *Sch* suprachiasmatic nucleus; *SL* lateral septal nucleus; *SM* medial septal nucleus; *Smt* submamillothalamic nucleus; *SN* substantia nigra; *Sth* subthalamic nucleus; *VB* Ventrobasal thalamic complex. Composed from Reinoso-Suárez and Llamas (1975); Reinoso-Suárez (1977); Pasquier and Reinoso-Suárez (1978)

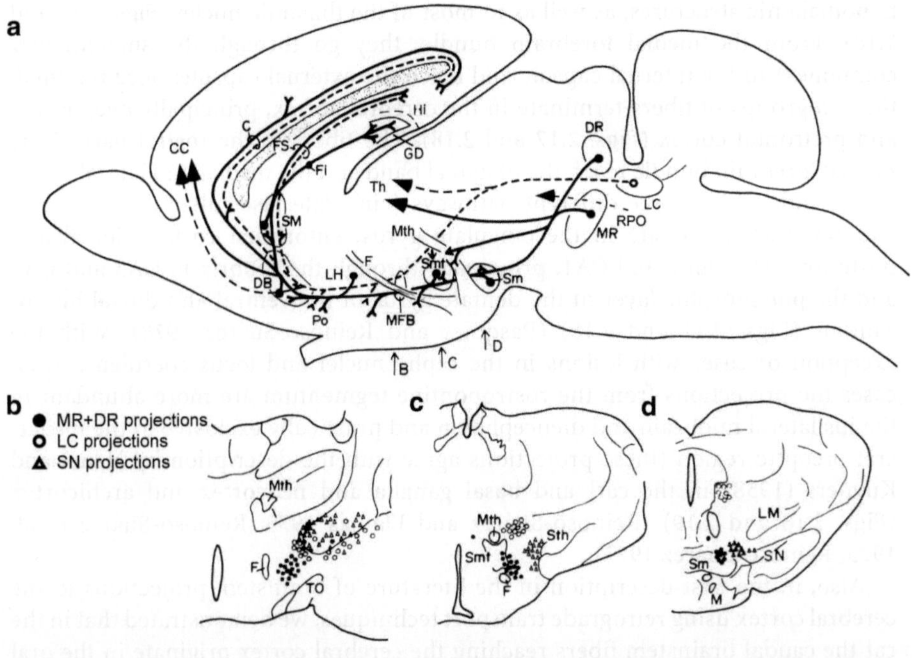

Fig. 2.18 Projections from the rostral pontine tegmentum and midbrain to hippocampus and cerebral cortex. (**a**) Scheme of a sagittal section of the rat brain showing the rostral pontine projections from the locus coeruleus in a broken line and from the raphe nuclei in a continuous line. Hypothalamic projections like those from the forebrain are represented joined to the raphe pathway. (**b**), (**c**), (**d**) Represent the main ascending pathways in three anterior- to posterior-coronal hypothalamic sections indicated by *arrows* in (**a**). *C* cingulate fasciculus; *CC* cerebral cortex; *DB* diagonal band (Broca); *DR* dorsalis raphe nucleus; *F* fornix; *FI* fimbria; *FS* fornix superior; *GD* dentate gyrus; *HI* hippocampus; *LC* locus coeruleus; *LH* lateral hypothalamic area; *MFB* medial forebrain bundle; *LM* medial lemniscus; *M* mamillar nucleus; *MR* medianus raphe nucleus; *Mth* mamillothalamic tract; *Po* preoptic region; *RPO* oral pontine reticular nucleus; *SM* medial septal nucleus; *Sm* supramamillary nucleus; *Smt* submamillothalamic nucleus; *SN* substantia nigra; *Sth* subthalamic nucleus; *Th* thalamus; *TO* optic tract. Modified from Pasquier and Reinoso-Suárez (1978) using dates from Reinoso-Suárez and Llamas (1975) and Reinoso-Suárez (1977)

degenerated fibers close to the midline as seen in the midbrain in addition to the previously mentioned degenerated fibers (Pasquier and Reinoso-Suárez 1978): (1) a dorsal group which ascends to the hypothalamus, passing medial to the mammillothalamic tract and joins the medial forebrain bundle in the dorsal hypothalamus [Fig. 2.17 (label 2)]; (2) a ventral group whose degenerated fibers run dorsal to the interpeduncular nucleus, medial in the ventral mesencephalic area, arriving at the posterior hypothalamus dorsolateral to the supramammillary nucleus and joining the medial forebrain bundle from a rostral direction [Fig. 2.17 (label 3)]. The two groups of fibers give collaterals and terminals to the midbrain and hypothalamic structures, as well as to most of the thalamic nuclei (Figs. 2.17 and 2.18). From the medial forebrain bundle, they go through the sublenticular component of the internal capsule and reach the external capsule; once reunited, the two groups of fibers terminate in the cerebral cortex, principally deep layers and prefrontal cortex (Figs. 2.17 and 2.18). The fibers in the rostral part of the medial forebrain bundle reach the diagonal band and the medial and lateral septal nuclei and, taking three different pathways (cingulate fasciculus, dorsal fornix, and fimbria), terminate in the cingulate gyrus, entorhinal cortex, dorsal and posterior subiculum, and CA1, principally through the fimbria in CA3 and CA4 and the polymorphic layer of the dentate gyrus of the ventral and dorsal hippocampus (Figs. 2.17 and 2.18) (Pasquier and Reinoso-Suárez 1978). With the exception of cases with lesions in the raphe nuclei and locus coeruleus, in all cases the projections from the rostropontine tegmentum are more abundant in the ipsilateral midbrain and diencephalon and practically exclusive to the ipsilateral preoptic region [these projections agree with the description by Nauta and Kuypers (1958) in the cat] and basal ganglia and neocortex and archicortex (Figs. 2.18 and 2.19) (Reinoso-Suárez and Llamas 1975; Reinoso-Suárez et al. 1975; Reinoso-Suárez 1977).

Also, in the first description in the literature of brainstem projections to the cerebral cortex using retrograde transport techniques, we demonstrated that in the cat the caudal brainstem fibers reaching the cerebral cortex originate in the oral pontine tegmentum, most numerously from the ipsilateral locus coeruleus complex, the parabrachial nuclei, and the raphe nuclei (centralis superior and raphe dorsalis); fibers originate less intensely from the oral pontine reticular nucleus and the contralateral corresponding structures since only a few neurons were observed in the latter (Fig. 2.20) (Llamas et al. 1975). Neurons in the same rostral pontine structures, with a topometric organization and a large participation of the raphe nuclei, were labeled after retrograde tracer injections in the cat and rat hippocampus (Fig. 2.21) (Pasquier and Reinoso-Suárez 1977, 1978). The oral pontine tegmentum is also the most caudal source of many thalamic projections from the brainstem reticular structures as revealed by retrograde tracer injections in the thalamus (Velayos and Reinoso-Suárez 1982; Rodrigo-Angulo and Reinoso-Suárez 1982, 1995; Jiménez-Castellanos and Reinoso-Suárez 1985).

All these projections connect the waking and REM sleep-controlling structures of the oral pontine tegmentum with the midbrain and forebrain structures related to SWC. The oral pontine tegmentum may exert actions directly on hypothalamic, thalamic, basal forebrain, and neocortica and archicortical formations more immediately than on the more caudal reticular structures within the same hemisphere; this would explain the preference by the ipsilateral side of the bioelectrical phenomena after a lesion or stimulation of structures situated in the oral pontine region in contrast with the more bilaterally balanced organization of the bioelectrical responses after lesion or stimulation of more caudal reticular formations (Figs. 2.18 and 2.19) (Reinoso-Suárez and Llamas 1975; Pasquier and Reinoso-Suárez 1978).

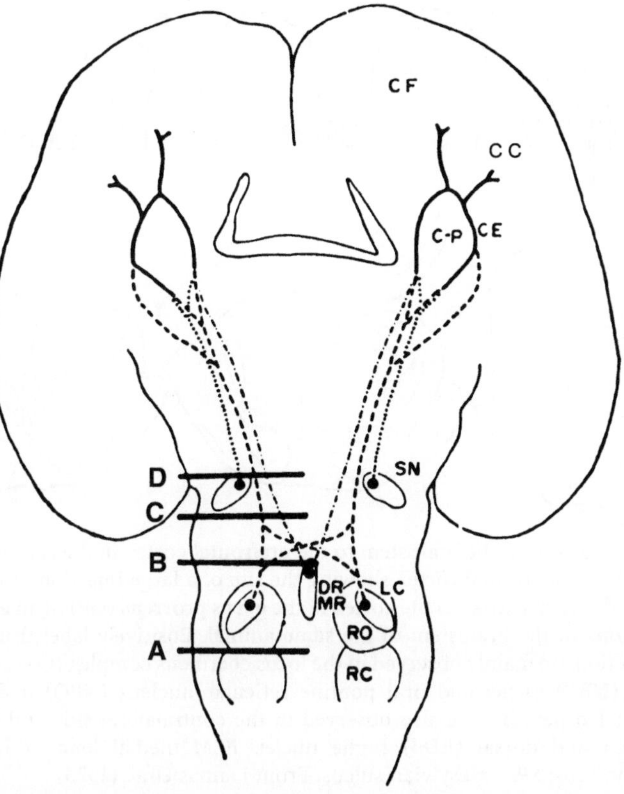

Fig. 2.19 Summary of the nuclei of origin and course of the projections from the brainstem to the cerebral cortex. (**a**) Mediopontine pretrigeminal hemisection, which does not produce degenerated fibers in the cerebral cortex. (**b**) Rostropontine pretrigeminal hemisection, which interrupted the projections from the locus coeruleus (LC) and other structures of the oral pontine tegmentum, like the parabrachial nuclei and the oral pontine reticular nucleus (RO), and destroyed the rostral raphe nuclei (DR and MR), gives rise to degenerated fibers in both cerebral cortices. (**c**) Caudal midbrain hemisection, rostral to the superior cerebellar peduncle decussation, produced degenerated fibers only in the side of the lesion; these fibers originate in locus coeruleus and parabrachial nuclei of both sides and the rostral raphe nuclei. (**d**) Hemisection rostral to the substantia nigra (SN): the projections from the ventral midbrain tegmentum joined the fibers interrupted in (**c**); thus, after a hemisection rostral to SN, degenerated fibers are seen only in the cerebral cortex on the side of the lesion. *CC* cerebral cortex; *CE* external capsule; *CF* frontal cortex; *C-P* caudatus-putamen. Modified from Reinoso-Suárez and Llamas (1975)

After midbrain and rostropontine hemisections (Fig. 2.10, Midbrain and Pons Rost.), degenerated terminal formations were observed in the oral and caudal pontine tegmentum, the medullary tegmentum and the spinal cord. These findings agree with descriptions by numerous authors (Torvik and Brodal 1957; Rossi

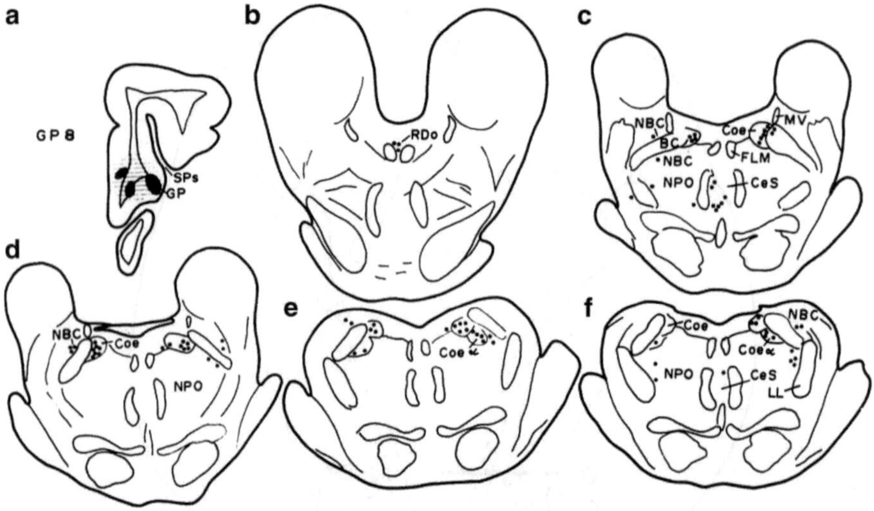

Fig. 2.20 Projections from the brainstem to the prefrontal cortex in the cat. (**a**) Scheme of a coronal section of the frontal cortex showing the site of a large injection of the retrograde tracer horseradish peroxidase in the cortex of the gyrus prorreus (GP) of an adult cat. (**b–f**) Coronal sections of the brainstem of the same animal. Positively labeled neurons (*dots*) after this injection are mainly observed in the locus coeruleus complex (Coe and Coe α) and parabrachial (NBC) nuclei and oral pontine reticular nucleus (NPO) ipsilateral to the injection. Labeled neurons are also observed in the contralateral side and in the central superior (CeS) and dorsal (RDo) raphe nuclei. *FLM* medial longitudinal fasciculus; *LL* lateral lemniscus; *SPs* praesylvian sulcus. From Llamas et al. (1975)

and Zanchetti 1957; Edwards 1975; Graybiel 1977; Sakai et al. 1979; Holstege and Kuypers 1982).

2.4 Midbrain Tegmentum, Hypothalamus, and Basal Forebrain

Villablanca (1965, 2004) reported the absence of REM sleep and presence of the usual polygraphic and pupil patterns that characterize both NREM sleep and W in the chronic *cerveau isolé* cat. These findings demonstrate that the midbrain and forebrain contains structures to organize these two states of the SWC. What structures are they?

Since the studies of Bremer (1935), the *midbrain* has been considered a basic structure for the regulation of the SWC with the mesencephalic tegmentum as the fundamental part of the ascending activating reticular system (Moruzzi and Magoun 1949; French and Magoun 1952). Lindsley et al. (1950) demonstrated that large lesions of the midbrain reticular formation that spared the classic ascending pathways produced EEG synchronization. They concluded that the

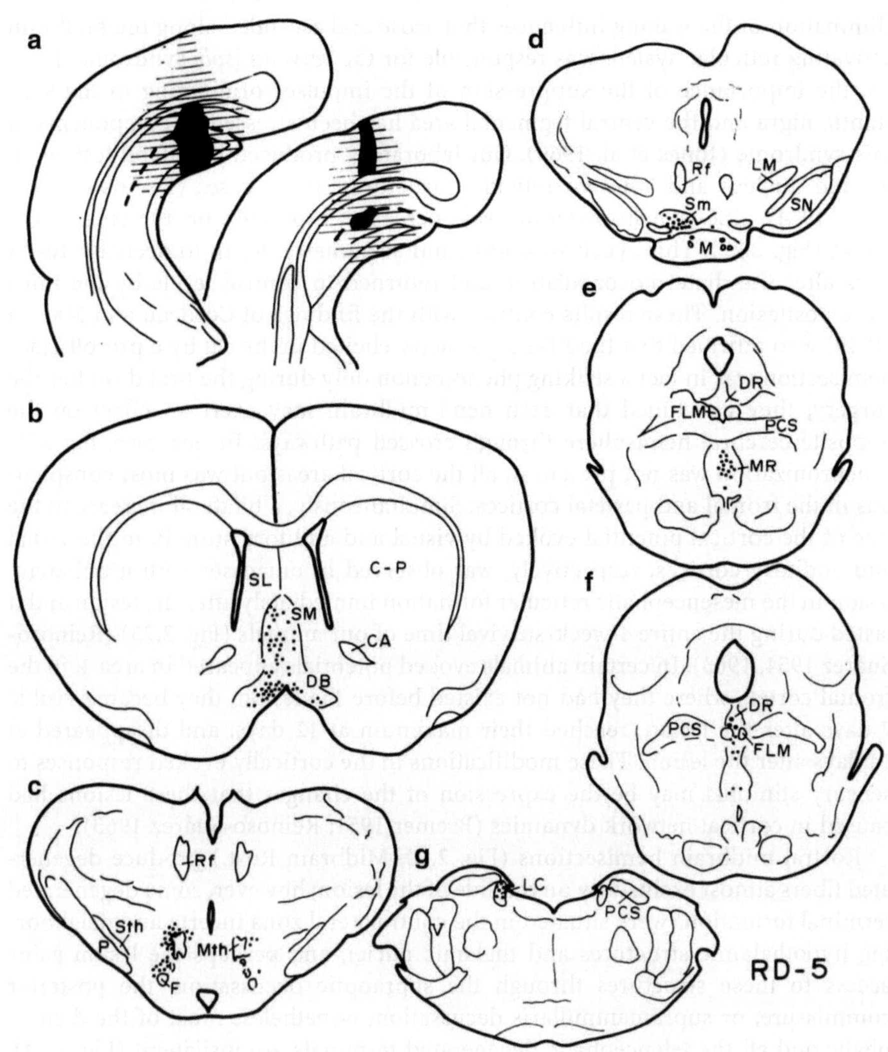

Fig. 2.21 Projections to the hippocampus in the rat. (**a**) Scheme of a coronal section of the left hemisphere showing the site of a large injection of the retrograde tracer horseradish peroxidase in the dorsal–posterior hippocampus of an adult rat. (**b–g**) Coronal sections of the brain of the same animal depicting positively labeled neurons (*dots*) after this injection; thalamic labeled neurons following neocortical peroxidase contamination are not represented. *C-P* caudatus-putamen; *CA* anterior commissure; *DB* diagonal band (Broca); *DR* dorsal raphe nucleus; *F* fornix; *FLM* medial longitudinal fasciculus; *LC* locus coeruleus; *LM* medial lemniscus; *M* mamillary nucleus; *MR* central superior raphe nucleus; *Mth* mamillothalamic tract; *P* pes pedunculi; *PCS* superior cerebellar peduncle; *Rf* fasciculus retroflexus; *SL* lateral septal nucleus; *SM* medial septal nucleus; *Sm* Supramamillar nucleus; *Smt* submamillothalamic nucleus; *SN* substantia nigra; *Sth* subthalamic nucleus; *V* sensory trigeminal nucleus. From Pasquier and Reinoso-Suárez (1978)

elimination of the waking influences that arose and ascended along the midbrain activating reticular system was responsible for the *cerveau isolé* syndrome. Later on, the importance of the suppression of the impulses originating in the substantia nigra and the ventral tegmental area has been stressed as components of this syndrome (Jones et al. 1969). Our laboratory produced unilateral lesions in the red nucleus and adjacent reticular structures. In all cases (Reinoso-Suárez 1952, 1954), EEG synchronization was increased, especially on the side of the lesion (Fig. 2.22). This synchronization and asymmetry began to decrease 10–14 days after the diatermocoagulation and returned to control levels by the third week postlesion. These results contrast with the findings of Cordeau and Mancia (1959) who affirmed that the EEG asymmetry elicited in the cat by a precollicular hemisection was in fact a striking phenomenon only during the first day after the surgery; they concluded that each hemi-midbrain may exert an effect on the opposite cerebral hemisphere through crossed pathways. In our case, the EEG synchronization was not present in all the cortical areas but was most conspicuous in the frontal and parietal cortices. Simultaneously, a bilateral increase in the size of the cortical potential evoked by visual and auditory stimuli, in the visual and auditory cortices, respectively, was observed in our cases with a unilateral lesion in the mesencephalic reticular formation immediately after the lesion and it lasted during the entire 4-week-survival time of our animals (Fig. 2.23) (Reinoso-Suárez 1954, 1963). In certain animals evoked potentials appeared in area 4, in the frontal cortex, where they had not existed before the lesion; they became visible 7 days after the lesion, reached their maximum at 12 days, and disappeared at 14 days after the lesion. These modifications in the cortically evoked responses to sensory stimulus may be the expression of the changes that these lesions had caused in cortical network dynamics (Bremer 1951; Reinoso-Suárez 1963).

Rostral midbrain hemisections (Fig. 2.10, Midbrain Rost.) produce degenerated fibers almost exclusively on the side of the lesion; however, some degenerated terminal formations were situated in the contralateral zona incerta and neighboring hypothalamic structures and thalamic nuclei, and perhaps the lesion gains access to these structures through the supraoptic decussation, the posterior commissure, or supramammillaris decussation; nonetheless most of the diencephalic and all the telencephalic degenerated terminals are ipsilateral (Fig. 2.24). Hemisections that destroy part of the substantia nigra and ventral tegmental area or are located rostral to these structures confirm the findings of Llamas (1966) and Llamas and Reinoso-Suárez (1969) of nigrothalamic, nigrostriate, and nigrocortical projections, with terminal degeneration appearing in the thalamus, subthalamic nucleus, globus pallidus, accumbens and caudate nuclei, olfactory tubercle, and cerebral cortex (Figs. 2.19, 2.24, and 2.25) (Reinoso-Suárez et al. 1975). This may explain the ipsi- and bilateral bioelectric effects of midbrain lesions, but we cannot exclude the reorganization of the cortical network dynamics across the corpus callosum. The importance of these diencephalic midline crossings and, principally, of the corpus callosum itself is easily understood by observing the behavior of the isolated hemicerebrum resulting from a sagittal midline section at

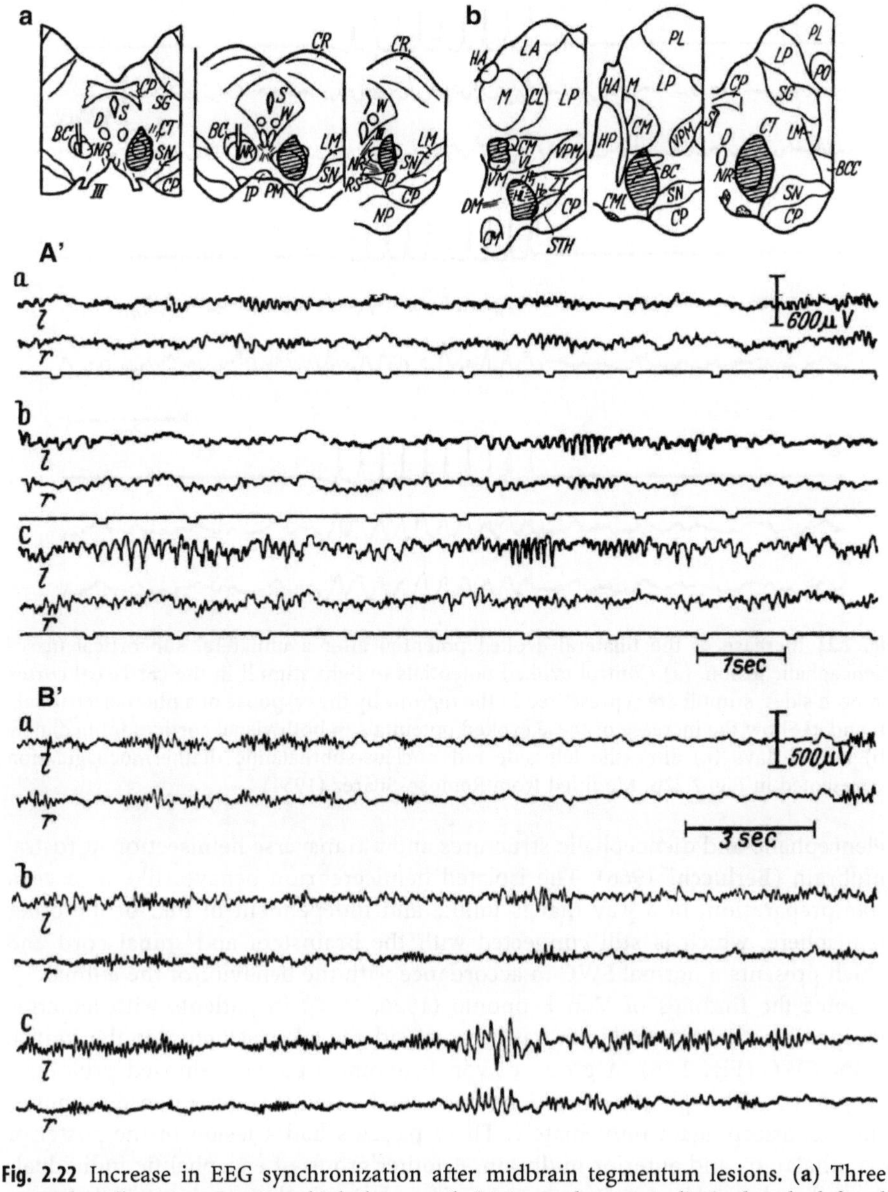

Fig. 2.22 Increase in EEG synchronization after midbrain tegmentum lesions. (**a**) Three coronal midbrain sections in which the coagulation spread was very limited to the left red nucleus (NR) and the brachium conjunctivum. As shown in (A′), this lesion produced an increase of EEG synchronization in the left area 5 (b, l) that began immediately after the lesion and was most intense 3 days later (c, l) in relation to the bilateral symmetry of the control (a, l, and r). (**b**) Three coronal sections showing the expanse of the lesion of the rostral part of the left red nucleus (NR) that extends rostral to the caudal subthalamus also damaging the brachium conjunctivum (BC), areas de Forel (H1 and H2), lateral hypothalamic area (HL), and extending to the ventromedial thalamic nucleus (VM). (**B′**) Shows the EEG records of left (l) and right (r) cat parietal cortices, before (a), immediately (b) and 5 day (c) after the lesion represented in (**b**). Observe the increased synchronizatin, especially in (c, l). Modified from Reinoso-Suárez (1954)

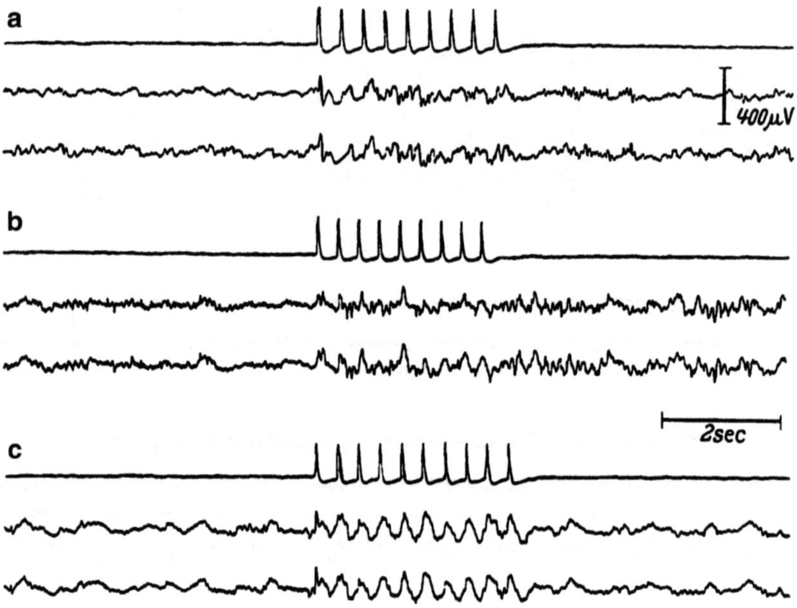

Fig. 2.23 Increase of the bilateral evoked potential after a unilateral subcortical meso-diencephalic lesion. (**a**) Control evoked potentials to light stimuli in the cat visual cortex on both sides, stimuli are represented in the *top row* by the response of a photoelectric cell. (**b** and **c**) Show the increase of these evoked potentials in both visual cortices immediately (b) and 5 days (c) after the left side red nucleus-subthalamic diathermocoagulation represented in Fig. 2.22b. Modified from Reinoso-Suárez (1954)

telencephalic and diencephalic structures and a transverse hemisection at rostral midbrain (Berlucchi 1966). The isolated hemicerebrum behaves like a *cerveau isolé* preparation, in a way that is unlike and independent of that of the other hemisphere, which is still connected with the brainstem and spinal cord and which presents a normal SWC in accordance with the behavior of the animal.

Since the findings of Von Economo (1926, 1930) in patients with lethargic encephalitis, the *hypothalamus* has been considered a key structure in the control of the SWC (Fig. 2.26). A group of von Economo's patients showed prolonged sleepiness. Although they could be awakened briefly by an intense stimulus, they fell asleep again immediately. These patients had a lesion in the posterior hypothalamus and anterior midbrain. Another group of encephalitic individuals showed prolonged insomnia accompanied by a lesion in the anterior hypothalamus. Based on these observations, von Economo (1930) suggested that the anterior hypothalamus was a sleep promoting structure, where a lesion would produce insomnia, and the posterior hypothalamus was a wakefulness promoting structure, where a lesion would increase sleep. Many experimental studies between the 1930s and 1970s confirmed this hypothesis in the last century: Ingram et al. (1936) in the cat, Ranson (1939) in the monkey, and Nauta (1946) in the rat showed that

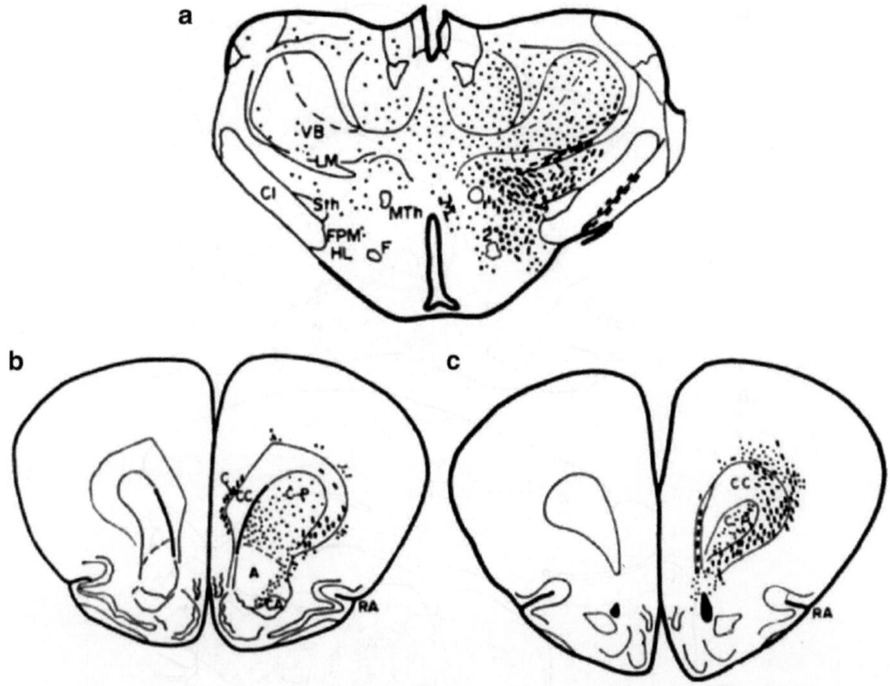

Fig. 2.24 Degenerated fibers after a midbrain rostral hemisection in the rat brain (see Fig. 2.10). (**a**, **b**, and **c**) Show, from caudal to rostral planes, the distribution, preferentially ipsilateral, of degenerated fibers in the rat diencephalon basal ganglia and rostral cerebral cortex after a midbrain rostral hemisection. *A* acumbens nucleus; *C* cingulum; *CA* anterior commissure; *CC* corpus callosum; *CI* internal capsule; *C-P* caudatus-putamen; *F* fornix; *FPM* medial forebrain bundle; *HL* lateral hypothalamic area; *LM* medial lemniscus; *Mth* mamillothalamic tract; *RA* anterior rhinal sulcus; *Sth* subthalamic nucleus; *VB* ventrobasal thalamic nucleus. Modified from Reinoso-Suárez and Llamas (1975)

the posterior and lateral hypothalamus is necessary to maintain W. Animals with lesions at that level showed a state of somnolence and indifference but although somnolent and indifferent, animals with these hypothalamic lesions could be momentarily awoken by a strong stimulus. In the early fifties, our group observed behavioral traits of stupidity and indifference in cats after bilateral lesions in the subthalamus and posterolateral hypothalamus (see Reinoso-Suárez and De Andrés 1976) similar to those described by Ranson (1939) in the monkey. We studied the changes occurring in the EEG after unilateral lesions in the diencephalon of cats; the lesions located in the lateral and posterior hypothalamus produced a widespread synchronization of the EEG in the same hemisphere [(Fig. 2.27) (Reinoso-Suárez 1959), see also Fig. 2.22b and B′]; in these animals, there was EEG synchronization in the frontal region even in periods of active behavioral W for 2 months postlesion (Fig. 2.28). Lesions in more rostral

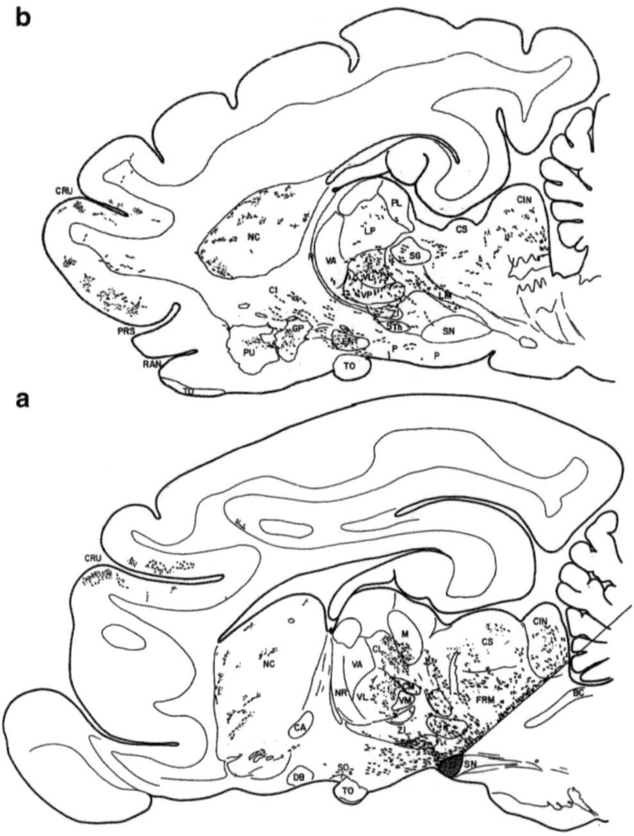

Fig. 2.25 Degenerated midbrain, diencephalic, basal forebrain, basal ganglia, and cortical fibers and terminals after a lesion of the medial part of the subtantia nigra and the ventral tegmental area. (**a**) Parasagittal section of the cat brain showing the lesion site and the consequently degenerated fibers. (**b**) Represents a more lateral parasagittal section showing the degenerated fibers (*interrupted lines*) and terminals (*dots*). The electrodes producing the lesions were placed in each animal following different approaches to rule out the degeneration produced by the electrode pathway. From Llamas and Reinoso-Suárez (1969)

areas, the lateral and dorsal hypothalamus as well as extending to the ventral border of the intralaminar and medial thalamic nuclei, generated synchronization circumscribed to anterior regions of the cortex (Fig. 2.27). Finally, lesions restricted to specific thalamic nuclei did not noticeably modify the EEG pattern.

As in the case of midbrain lesions, unilateral subthalamic–hypothalamic lesions also produced bilateral increased evoked potentials after auditory and visual stimuli, even in cortical areas in which evoked potentials had not previously been recorded in nonlesioned animals (cats and rabbits) (Figs. 2.29 and 2.30) (Reinoso-Suárez 1960, 1962, 1963); see also Figs. 2.22 and 2.23 (Reinoso-Suárez

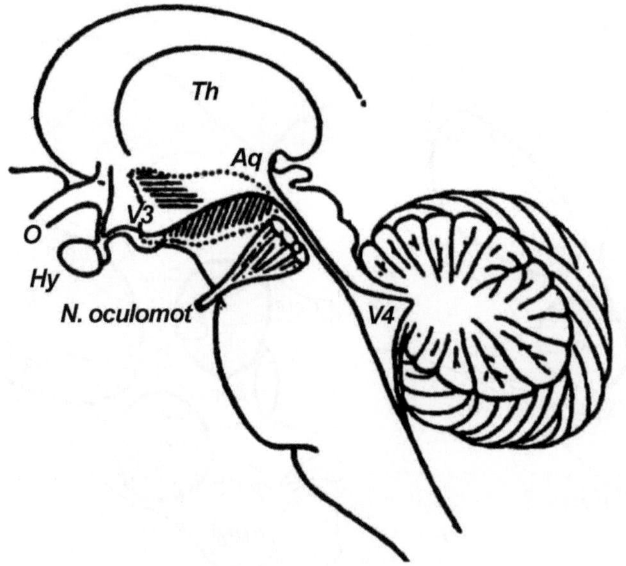

Fig. 2.26 von Economo's illustration of the hypothalamic areas (limited by *dots*) in which lesion produces modifications in the SWC. Diagonal hatching, posterior hypothalamus-midbrain junction where a lesion causes prolonged sleepiness; horizontal hatching, anterior hypothalamus where a lesion produces insomnia. *Aq* cerebral aqueduct; *Hy* hypophysis; *O* optic nerve; *Th* thalamus; *V3* third ventricle; *V4* fourth ventricle

1954). The diathermocoagulations that increase the evoked potential response in cats to auditory stimuli in the auditory cortex ipsi- and contralaterally to the coagulation lesion are those located in the medial half of the zona incerta, neighboring dorsal, posterior and lateral hypothalamus, and medial and ventral nuclei of the dorsal thalamus, such as the ventromedial, centromedial, and submedial thalamic nuclei (Fig. 2.29). Evoked potentials may be recorded after a lesion in other cerebral areas like the frontal cortex in some cases. Coagulation of the medial geniculate nucleus produced a decrease of auditory evoked potentials in the auditory cortex. Lesion of the remaining thalamic nuclei or the hypothalamic and dorsomedial thalamic structures situated in front of the zona incerta did not alter the evoked potentials to acoustic stimuli (Fig. 2.29). In rabbits with similar subthalamic–hypothalamic lesions, we found a bilateral increase of the primary evoked potential response to single flashes of light in the visual cortices (experiments done in collaboration with Joaquín Fuster). These increases lasted the 2 weeks of the experiments (Fig. 2.30) (Reinoso-Suárez 1963). The decreased evoked potential response to visual stimuli also observed during the 2 weeks following lesion in the lateral geniculate nucleus demonstrated that the increase of the evoked potentials in the cerebral cortex depended, as we had suggested previously, on phenomena that occur in the dynamic network of the cerebral cortex (Fig. 2.30) (Reinoso-Suárez 1954, 1963).

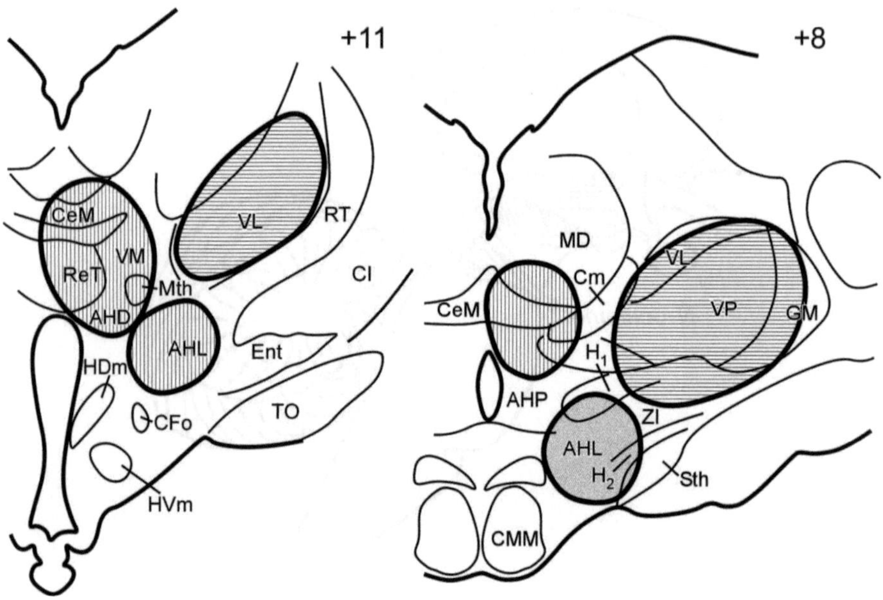

Fig. 2.27 Diencephalic lesions that produce EEG synchronization in the ipsilateral hemisphere. Diagram of the diencephalic part of planes +8 and +11 of the stereotaxic cat atlas by Reinoso-Suárez (1961), showing the location of unilateral lesions (*dotted area*), produced by diatermocoagulation, which produce a generalized synchronization of the EEG in the same hemisphere as the lesion. Those lesions that gave rise to less defined modifications limited to the anterior cortical areas are shown in vertical lines. The areas labeled by horizontal lines represent lesions that did not produce modifications in the EEG. AHD dorsal hypothalamic area; AHL lateral hypothalamic area; AHP dorsal hypothalamic area; CeM central medial nucleus; CFo fornix; CI internal capsel; CMM medial mamilar nucleus; Cm centromedian nucleus; Ent entopeduncular nucleus; GM medial geniculate nucleus; H_1 H_1 area; H_2 H_2 area; HDm dorsomedial hypothalamic nucleus; HVm ventromedial hypothalamic nucleus; MD mediodorsal nucleus; Mth mamillothalamic tract; ReT reuniens nucleus; RT reticular thalamic nucleus; Sth subthalamic nucleus; TO optic tract; *VL* ventral lateral nucleus; *VM* ventral medial nucleus; *VP* ventral posterior nucleus; *ZI* zona incerta. Composed from Reinoso-Suárez (1959)

These modifications in bioelectric and behavioral patterns after hypothalamic lesions may, at least in part, be caused by damage to those nerve fibers that, having originated in more caudal levels, are just passing through this region. However, conspicuous projections from these hypothalamic regions to the thalamus, cerebral cortex, hippocampus, and basal forebrain as well as caudal brainstem regions guarantee a direct participation by the hypothalamus in these behavioral and bioelectric phenomena (Reinoso-Suárez 1963, 1977, 1985; Pasquier and Reinoso-Suárez 1976, 1978; Nauta and Haymaker 1969; Reinoso-Suárez and De Andrés 1976; Velayos and Reinoso-Suárez 1982; Avendaño and Llamas 1983; Jiménez-Castellanos and Reinoso-Suárez 1985; Clascá et al. 1989).

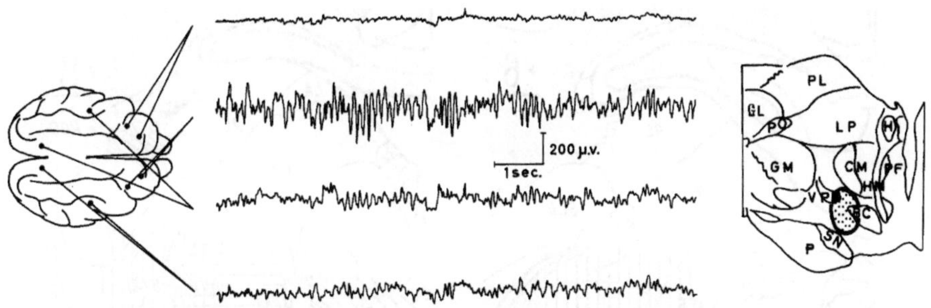

Fig. 2.28 EEG synchronization after subthalamic diathermocoagulation. The figure (Reinoso-Suárez and De Andrés 1976) shows a period of EEG recording from a free-moving cat with implanted electrodes, in a situation of active W, 48 days after the lesion in the right posterior subthalamic region. Observe the EEG synchronization in the right frontal record. *BC* brachium conjunctivum; *CM* centromedian nucleus; *GL* lateral geniculate nucleus; *GM* medial geniculate nucleus; *H* habenular nucleus; *HM* retroflex bundle; *LP* lateral posterior nucleus; *P* cerebral peduncle; *PF* parafascicular nucleus; *PO* Posterior nucleus; *SN* substantia nigra; *VP* ventral posterior nucleus. Composed from Reinoso-Suárez (1959)

von Economo (1930) suggested that the anterior hypothalamus was a sleep promoting structure, one in which lesion would produce insomnia (Fig. 2.26), and the findings of Hess (1931). Nauta (1946), Sterman and Clemente (1962) and Hernández-Peón (1962) from stimulation or lesions in the *basal forebrain* and *anterior hypothalamic regions* encouraged us to explore the role that these regions play in regulating the SWC in the cat. Earlier, in 1954, in unpublished experiments in cats, we had observed that lesions of the diagonal band of Broca increased EEG activation, accompanied by a behavioral increase of active W, with decreased EEG synchronization. Madoz (1968) showed that small diatermocoagulation lesions, both uni- and bilateral, in the preoptic region, in the horizontal limb of the diagonal band of Broca, in the nucleus accumbens, and in the adjacent very ventral part of the head of the caudate nucleus could decrease sleep, mainly NREM sleep, with an associated increase in W and, to a lesser degree, also drowsiness (Fig. 2.31). These findings proved the participation of the aforementioned structures in a rostral hypnogenic system (Madoz and Reinoso-Suárez 1968) that McGinty and Sterman (1968) situated more caudally after ample bilateral lesions extending to the anterior hypothalamus. Basal forebrain connections with cerebral cortex, hippocampus (Figs. 2.18 and 2.21), thalamus, hypothalamus, and brainstem are considered the morphological support for basal forebrain participation in the organization of SWC sleep phases (Nauta and Haymaker 1969; Divac 1975; Kievit and Kuypers 1975a, b; Pasquier and Reinoso-Suárez 1976, 1978; Reinoso-Suárez et al. 1982; Avendaño and Llamas 1983; Reinoso-Suárez 1985; Velayos and Reinoso-Suárez 1985; Steriade et al. 1987a).

The interconnections between the sleep promoting structures of the basal forebrain-anterior hypothalamus and the waking structures in the rostral

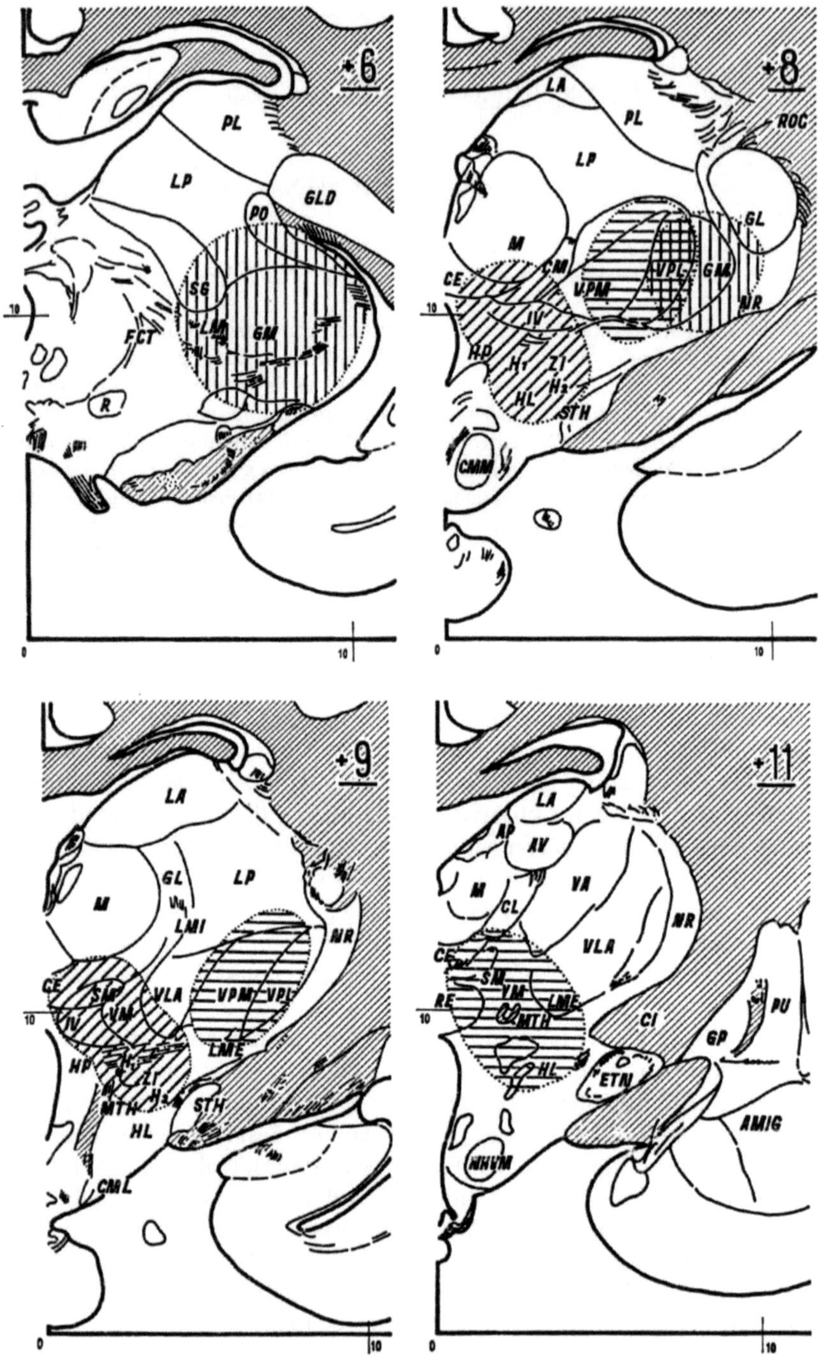
+6
PL
LP
GLD
PO
SG
LM
GM
FCT
R
10
0
+8
LA
PL
ROC
LP
GL
M
CM
VPL
GM
CE
VPM
MR
IV
HP
ZI
HL
STH
CMM
+9
LA
GL
LP
M
LMI
MR
CE
SM
VLA
VPM
VPL
IV
VM
LME
HP
ZI
MTH
STH
HL
CML
+11
LA
AP
AV
M
VA
CL
MR
VLA
CE
SM
RE
VM
MTH
LME
CI
PU
GP
HL
ETN
AMIG
NHVM

brainstem-lateralposterior hypothalamus are noteworthy. For example, Bremer (1970) demonstrated that the mesencephalic reticular formation, as part of the waking system, may be inhibited by impulses arising in the basal forebrain. This result shows that the anterior hypothalamic and preoptic regions are sleep-generating structures that interact with the waking formations of the posterior lateral hypothalamus and rostral brainstem, and that they constitute the primary forebrain network responsible for the organization of the SWC (Reinoso-Suárez and De Andrés 1976). Both systems would be modulated by the excitatory or inhibitory impulses coming from the caudal brain-stem and rostral telencephalic structures (Fig. 2.32).

2.5 Thalamus

Although restricted lesions in most of the thalamic nuclei do not produce conspicuous modifications in the EEG pattern, today we know that the *thalamus*, the other diencephalic structure, plays an important role in the organization of the SWC. In our experience, only lesions in the lateral and dorsal hypothalamus that extend to the ventral border of the intralaminar and medial thalamic nuclei generate EEG synchronization, and that this synchronization circumscribed to anterior regions of the cortex (Fig. 2.27) (Reinoso-Suárez 1959). Morrison and Dempsey (1942) showed that the thalamus is necessary for the generation of both electrophysiological recruiting responses and their counterpart the physiological

Fig. 2.29 Diencephalic areas related with the modification of the evoked potential responses to auditory stimuli. The figure is a schematic representation of the diencephalic part of the frontal sections at the levels +6, +8, +9, and +11 from the cat sterotaxic atlas by Reinoso-Suárez (1961) showing the areas where diathermocoagulation modifies the amplitude of the evoked potential responses to auditory stimulus. *Oblique lines*: coagulations which result in an increase of the evoked potential; *vertical lines*: those which produce a decrease; *horizontal lines*: those which do not alter the evoked potentials to acoustic stimuli. *AD* anterior dorsal nucleus; *AV* anterior ventral nucleus: *CE* central medial nucleus; *CI* internal capsule; *CL* central lateral nucleus; *CM* centromedian nucleus; *CML* lateral mammillary nucleus; *CMM* medial mammillary nucleus; *ENT* entopeduncular nucleus; *FCT* tegmental central fascicle; *GL* lateral geniculate nucleus; *GM* medial geniculate nucleus; *GP* globus pallidus; *H1* and *H2* Forel fields; *HL* lateral hypothalamic area; *HP* posterior hypothalamic area; *LA* lateral anterior nucleus; *LM* medial lemniscus; *LME* external medullary lamina; *LMI* internal medullary lamina; *LP* lateral posterior nucleus; *M* mediodorsal nucleus; *MTH* mammillothalamic tract; *NHVM* ventromedial hypothalamic nucleus; *NR* reticular nucleus; *PL* pulvinar nucleus; *PO* posterior nucleus; *R* red nucleus; *RE* reuniens nucleus; *SG* suprageniculate nucleus; *SM* submedial nucleus; *STH* subthalamic nucleus; *VA* ventral anterior nucleus; *VLA* ventral lateral nucleus; *VM* ventral medial nucleus; *VPL* ventral posterolateral nucleus; *VPM* ventral posteromedial nucleus; *ZI* zona incerta. From Reinoso-Suárez (1960)

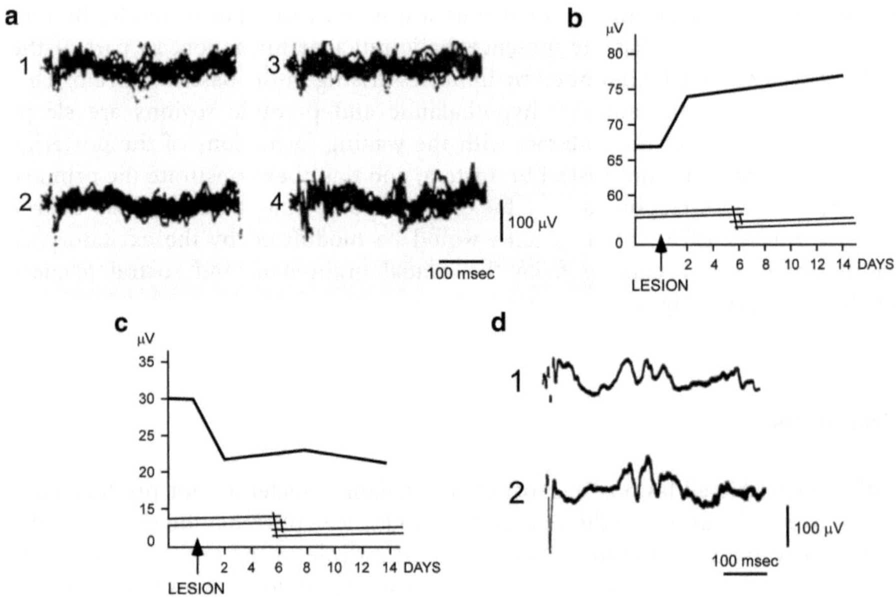

Fig. 2.30 Modification of the evoked potentials to visual stimuli in rabbit visual cortex after unilateral subthalamic–hypothalamic lesions (experiment done in collaboration with Joaquín Fuster). (**a**) Evoked responses in visual cortex to visual stimuli (stroboscopic flash); each strip shows ten superimposed traces; (1) before subcortical lesion, (2) 2 days after lesion, (3) 1 week after lesion, (4) 2 weeks after lesion. (**b**) Average of the increase in primary components of the evoked potentials in the rabbit's visual cortex to visual stimuli after unilateral subthalamic–hypothalamic lesions. (**c**) Average of the decrease in the evoked response in the rabbit lateral geniculate body to visual stimuli after subthalamic–hypothalamic lesion. (**d**) Increase of the evoked responses in visual rabbit cortex to ipsilateral geniculate shock after unilateral subthalamic–hypothalamic lesions; (1) before lesion, (2) 2 weeks after lesion. Modified from Reinoso-Suárez (1963)

EEG sleep spindles that characterize phase 2 of NREM sleep. Previously, Hess (1927, 1968) demonstrated that low-frequency electrical stimulation of the thalamic region located lateral to the midline thalamic nuclei produced a progressive change from W to a genuine, well-integrated physiological sleep pattern. Moruzzi supposed that so complex a phenomenon as sleep might be triggered by electrical stimulation of the thalamus, since the thalamus would hold the final common path of all the neural mechanisms actively leading to sleep (Moruzzi 1972). Villablanca (1974) demonstrated the absence of sleep spindles in athalamic cats in which the behavioral postures of sleep were also absent. In the first days after the thalamus ablation, the athalamic animals showed insomnia and behavioral hyperactivity together with an EEG with delta and slow activity, and low voltage fast W rhythms only appeared later on. The chronic athalamic cat shows W and REM sleep but not typical NREM sleep (Villablanca 2004). Steriade et al. (1987a, b)

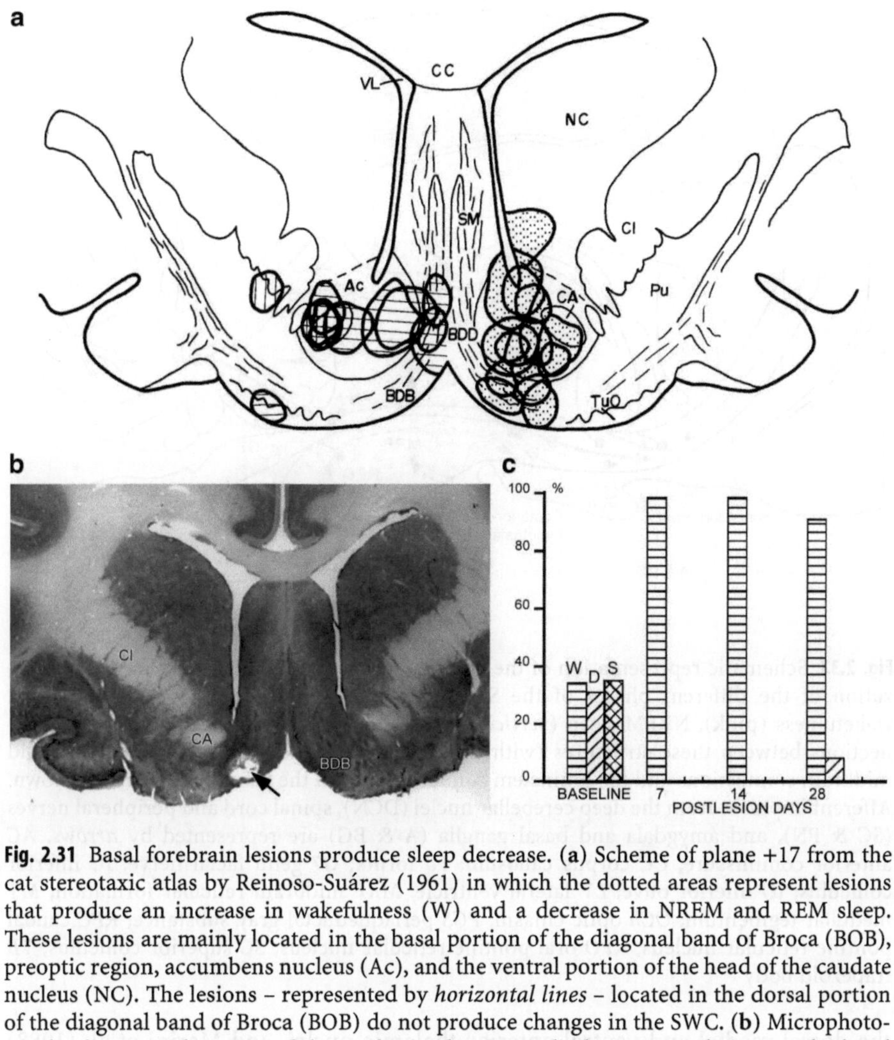

Fig. 2.31 Basal forebrain lesions produce sleep decrease. (**a**) Scheme of plane +17 from the cat stereotaxic atlas by Reinoso-Suárez (1961) in which the dotted areas represent lesions that produce an increase in wakefulness (W) and a decrease in NREM and REM sleep. These lesions are mainly located in the basal portion of the diagonal band of Broca (BOB), preoptic region, accumbens nucleus (Ac), and the ventral portion of the head of the caudate nucleus (NC). The lesions – represented by *horizontal lines* – located in the dorsal portion of the diagonal band of Broca (BOB) do not produce changes in the SWC. (**b**) Microphotograph of a coronal section of the cat brain showing a diathermocoagulation on the basal portion of the diagonal band of Broca (*arrow*) that produced the increase in W and the decrease in sleep shown in (**c**); these results were obtained during 2-h recordings, which were uniform in time and temperature over the 3 weeks previous to the lesion – basal – and 4 weeks postlesion. *CA* anterior commissure; *CI* internal capsule; *D* drowsiness; *Pu* putamen; *S* sleep (NREM and REM sleep); *SM* medial septal nucleus; *TuO* olfactory tubercle. Composed from Madoz (1968) and Madoz and Reinoso-Suárez (1968)

demonstrated the necessary contribution of the reticular thalamic nucleus as a pacemaker for sleep spindles. Sforza et al. (1995) showed that the pathognomonic lesion in the "fatal familial insomnia" syndrome (Lugaresi et al. 1986), which courses with a dramatic reduction of sleep (specially of NREM sleep), is injury of

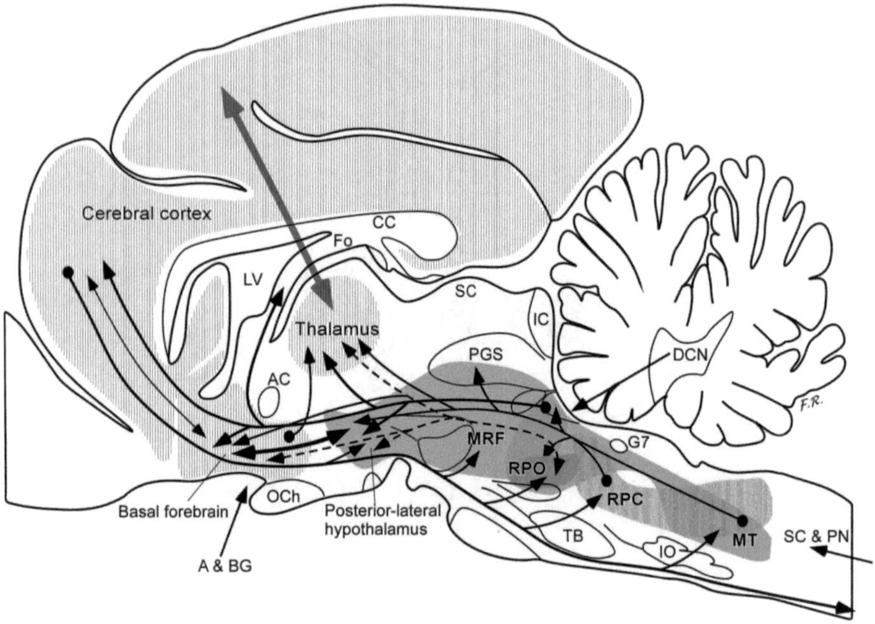

Fig. 2.32 Schematic representation of the encephalic structures responsible for the organization of the different phases of the SWC in a parasagittal section of the cat brain: wakefulness (pink), NREM sleep (*vertical lines*), and REM sleep (green). The main connections between these structures (with the exception of the efferent hypothalamic and midbrain connections and the brainstem connections from the basal forebrain) are shown. Afferent impulses from the deep cerebellar nuclei (DCN), spinal cord and peripheral nerves (SC & PN), and amygdala and basal ganglia (A & BG) are represented by *arrows*. *AC* anterior commissure; *CC* corpus callosum; *Fo* fornix; *G7* genu facial nerve; *IC* inferior colliculus; *IO* inferior olive; *LV* lateral ventricle; *MRF* midbrain reticular formation; *MT* medullar tegmentum; *OCh* optic chiasm; *PGS* periaqueductal gray substance; *RPC* caudal pontine reticular nucleus; *RPO* oral pontine reticular nucleus; *SC* superior colliculus; *TB* trapezoid body

the dorsal medial and ventral anterior thalamic nuclei, and Marini et al. (1988) demonstrated that experimental lesions of the dorsal medial thalamic nucleus, essentially its intermediate portion, decrease sleep, mainly NREM sleep, in the cat.

The thalamus in mammals is a sizeable and complex structure constituted by a large number of nuclei. We may schematize these nuclei as: midline, medial, intralaminar, lateral, ventral, and reticular (Fig. 2.33). Most of the nuclei, specially the midline, dorsal medial, intralaminar, and reticular thalamic nuclei, receive ample afferents from the brainstem and hypothalamic and basal forebrain structures related with the control of the SWC [Figs. 2.32, 2.33 (gray transparent arrows), 2.34, and 2.35]. The cerebral cortex is not only the principal source of thalamic afferents but also the principal efferent target of thalamic nuclei with the exception of the projections to the basal ganglia from the "intralaminar" and

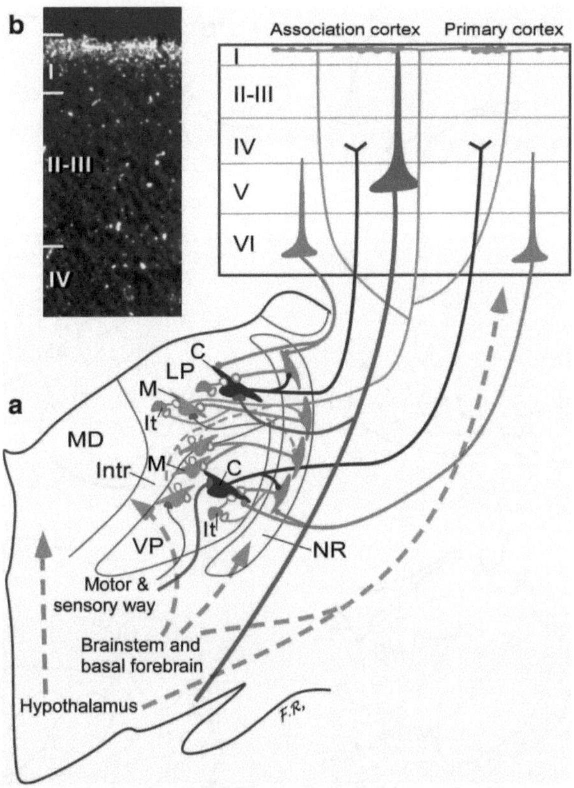

Fig. 2.33 Thalamic nuclei, cell types, and thalamocortical relationships. (**a**) The bottom represents a schematic coronal section of the primate diencephalon and the top a section of the primary and association cerebral cortices. The medial dorsal (MD), lateral posterior (LP), and ventral posterior (VP) nuclei are represented as examples of the medial, lateral, and ventral nuclei, respectively. *C* core type cells; *Intr* intralaminar thalamic nuclei; *It* GABAergic interneurons; *M* matrix type cells; *NR* reticular thalamic nucleus; *I-VI* cortical layers. The projections of the reticular thalamic nucleus neurons to the thalamic nuclei interneurons are not represented. (**b**) Coronal section at the level of the cat prefrontal cortex showing intense labeling in the superficial sector of layer I after an injection of an anterograde tracer in ventral medial nucleus, a representative site for matrix-type cells

closely related nuclei; finally, fibers originating from either the reticular thalamic nucleus or, to a more limited extent, from the posterior intralaminar nuclei terminate in the deep layers of the superior colliculus, periaqueductal gray matter, and midbrain and oral pontine reticular nuclei (Grofová et al. 1978; Tortelly and Reinoso-Suárez 1980; Parent and Steriade 1984; Reinoso-Suárez et al. 1990, 1994).

The thalamus contains interneurons and thalamocortical projection neurons. All the interneurons are GABAergic; carnivores and primates have two types of interneurons: the reticular nucleus neurons present in all mammals and the

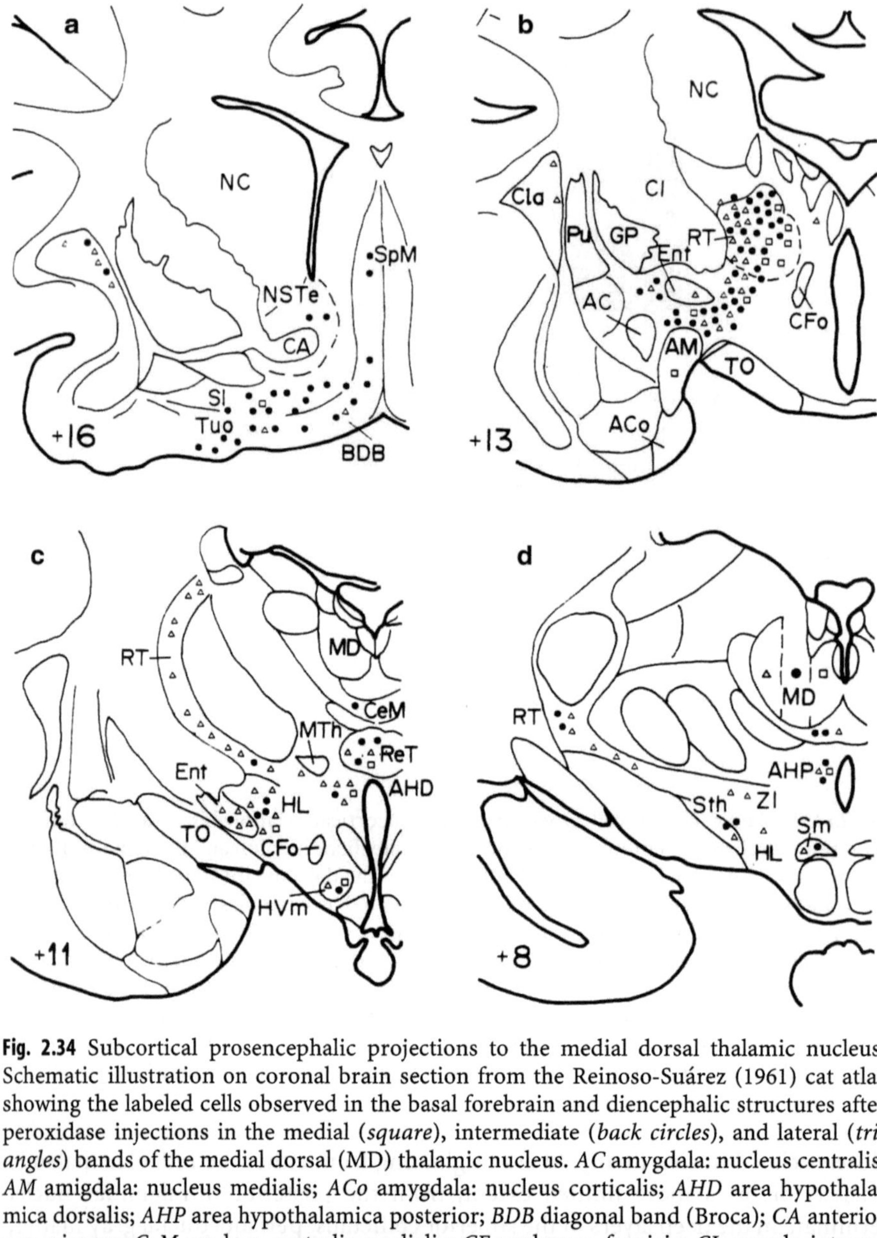

Fig. 2.34 Subcortical prosencephalic projections to the medial dorsal thalamic nucleus. Schematic illustration on coronal brain section from the Reinoso-Suárez (1961) cat atlas showing the labeled cells observed in the basal forebrain and diencephalic structures after peroxidase injections in the medial (*square*), intermediate (*back circles*), and lateral (*triangles*) bands of the medial dorsal (MD) thalamic nucleus. *AC* amygdala: nucleus centralis; *AM* amigdala: nucleus medialis; *ACo* amygdala: nucleus corticalis; *AHD* area hypothalamica dorsalis; *AHP* area hypothalamica posterior; *BDB* diagonal band (Broca); *CA* anterior commissure; *CeM* nucleus centralis medialis; *CFo* columna fornicis; *CI* capsula interna; *Cla* claustrum; *Ent* nucleus entopeduncularis; *GP* globus palidus; *HL* area hypothalamica lateralis; *HVm* area hypothalamica ventromedialis; *Mth* fasciculus mamillothalamicus; *MD* nucleus medialis dorsalis; *Mth* fasciculus; mamillothalamicus; *NC* nucleus caudatus; *NSTe* nucleus striae terminalis; *Pu* putamen; *ReT* nucleus reuniens thalami; *RT* reticular thalamic nucleus; *SI* substantia innominata; *Sm* nucleus submedius; *Spm* nucleus septalis medialis; *Sth* nucleus subthalamicus; *TO* tractus opticus; *TuO* tuberculum olfatorium; *ZI* zona incerta. From Velayos and Reinoso-Suárez (1985)

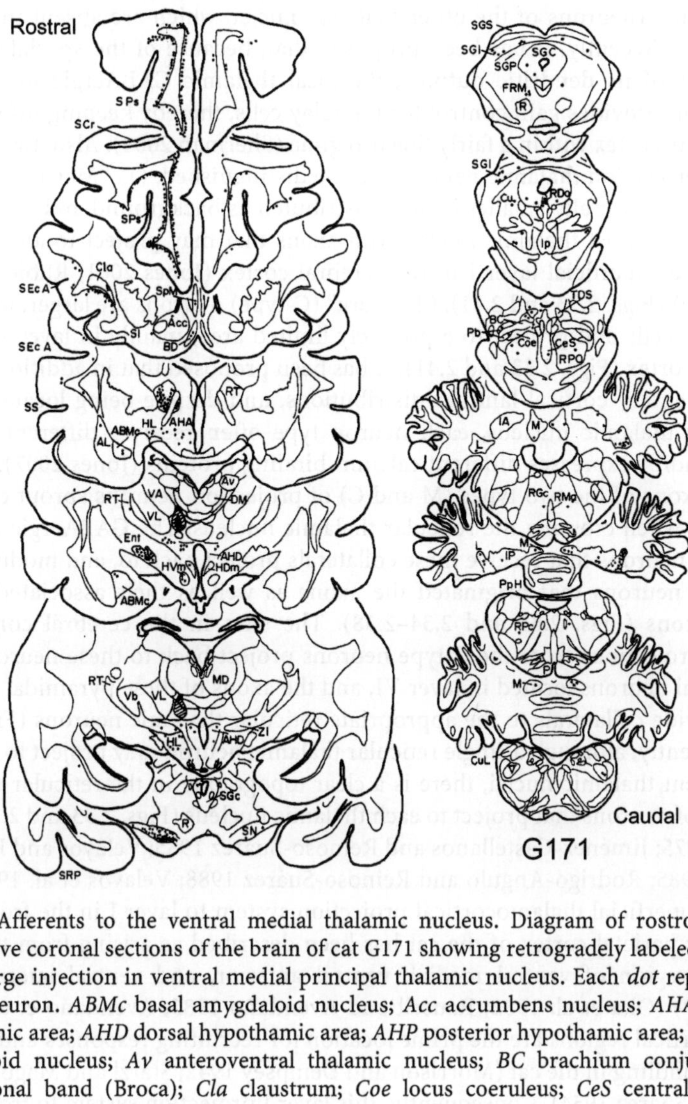

Fig. 2.35 Afferents to the ventral medial thalamic nucleus. Diagram of rostro-caudally consecutive coronal sections of the brain of cat G171 showing retrogradely labeled neurons after a large injection in ventral medial principal thalamic nucleus. Each *dot* represents a labeled neuron. *ABMc* basal amygdaloid nucleus; *Acc* accumbens nucleus; *AHA* anterior hypothamic area; *AHD* dorsal hypothamic area; *AHP* posterior hypothamic area; *AL* lateral amygdaloid nucleus; *Av* anteroventral thalamic nucleus; *BC* brachium conjunctivum; *BD* diagonal band (Broca); *Cla* claustrum; *Coe* locus coeruleus; *CeS* central superior raphe nucleus; *Cu* cuneiform nucleus; *CuL* lateral cuneatus nucleus; *DM* mediodorsal thalamic nucleus; *Ent* entopeduncular nucleus; *FRM* mesencephalic reticular formation; *HL* lateral hypothalamic area; *HDm* dorsomedial hypothalamic nucleus; *HVm* ventromedial hypothalamic nucleus; *IA* anterior interposed cerebellar nucleus; *IP* posterior interposed cerebellar nucleus; *Ip* interpeduncular nucleus; *L* lateral cerebellar nucleus; MD mediodorsal thalamic nucleus; *P* Pyramidal tract; *Pb* parabrachial nucleus; *PpH* nucleus praepositus hypoglossi; *R* red nucleus; *RDo* dorsal raphe nucleus; *RGc* gigantocelular reticular nucleus; *RMg* nucleus raphe magnus; *RPc* parvicellular reticular nucleus; *RPO* oral pontine reticular nucleus; *RT* reticular thalamic nucleus; *SCr* cruciate sulcus; *SGC* central gray matter; *SEcA* anterior ectosylvianus sulcus; *SGI* superior colliculus intermediate layer; *SGP* superior colliculus deep layer; *SGS* superior colliculus superficial layer;

intrinsic interneurons of the other thalamic nuclei, which are absent in rodents (Fig. 2.33). Recently, it has been proposed that, because of the special synaptic properties of its dendritic outputs, the local thalamic GABAergic interneuron population provides gain control for the relay cells, thereby keeping information relay to the cortex within a fairly linear regime (Sherman 2004). Also, two types of cortical projection thalamic neurons can be distinguished: (1) "Matrix" (M-type) neurons are generally smaller in size, distributed as background and constitute a large part of most thalamic nuclei, with axons that may project to an extensive area at the superficial layer I of the cerebral cortex (Jones 2001; Rubio-Garrido et al. 2009) (Figs. 2.33 and 2.41). (2) "Core" (C-type) neurons are larger, form core groups of cells and project to a precisely limited area, mainly in layer IV of the cerebral cortex (Figs. 2.33 and 2.41). It has been proposed that in addition to their complementary cortical laminar distributions, and despite being located within the same thalamic nucleus, each neuron type often receives different afferent inputs and/or expresses different calcium-binding proteins (Jones 2007).

The axons of the two types (M and C) of projection neurons sprout collateral branches when crossing the reticular thalamic nucleus. The GABAergic reticular thalamic neurons that receive these collaterals project back to, and modulate, the thalamic neurons that originated the axons as well as their associated related interneurons (Figs. 2.33 and 2.34–2.38). The sites in the cerebral cortex that receive projections from the C-type neurons project back to these neurons from pyramidal neurons located in layer VI, and the axons of these pyramidal neurons also provide collaterals to the appropriate reticular thalamic neurons (Fig. 2.33). Consequently, although a single reticular thalamic neuron may project to neurons in different thalamic nuclei, there is a clear topography in the reticular thalamic nucleus of neurons that project to each thalamic nucleus (Figs. 2.33 and 2.34–2.38) (Jones 1975; Jiménez-Castellanos and Reinoso-Suárez 1985; Velayos and Reinoso-Suárez 1985; Rodrigo-Angulo and Reinoso-Suárez 1988; Velayos et al. 1989).

The superficial thalamocortical projection system to layer I in the frontal and posterior parietal cortex of the cat has been described as arising from the paralaminar region of ventral medial, ventral anterior, and ventral lateral nuclei (Fig. 2.39) (Oka et al. 1982; Rausell and Avendaño 1985; Avendaño et al. 1990). These cortical regions are the prime location for recruiting responses and spontaneous spindling in the cat (Morrison and Dempsey 1942; Starzl and Magoun 1951; Reinoso-Suárez 1954). Consequently, this layer I projection system to frontal and posterior parietal cortices may be the final path for recruiting responses and spontaneous spindling activities (Oka et al. 1982; Avendaño et al. 1990).

Fig. 2.35 (continued) SI substantia innominata; *SN* substantia nigra; *SRP* posterior rhinal sulcus; *SS* sylvian sulcus; *SPs* presylvian sulcus; *SpM* medial septal nucleus;*TDS* dorsal tegmental nucleus (Gudden); *V* spinal trigeminal nucleus; *VA* ventral anterior thalamic nucleus; *Vb* ventrobasal complex; *VL* ventral lateral thalamic nucleus; *ZI* zona incerta. Modified from Jiménez-Castellanos and Reinoso-Suárez (1985)

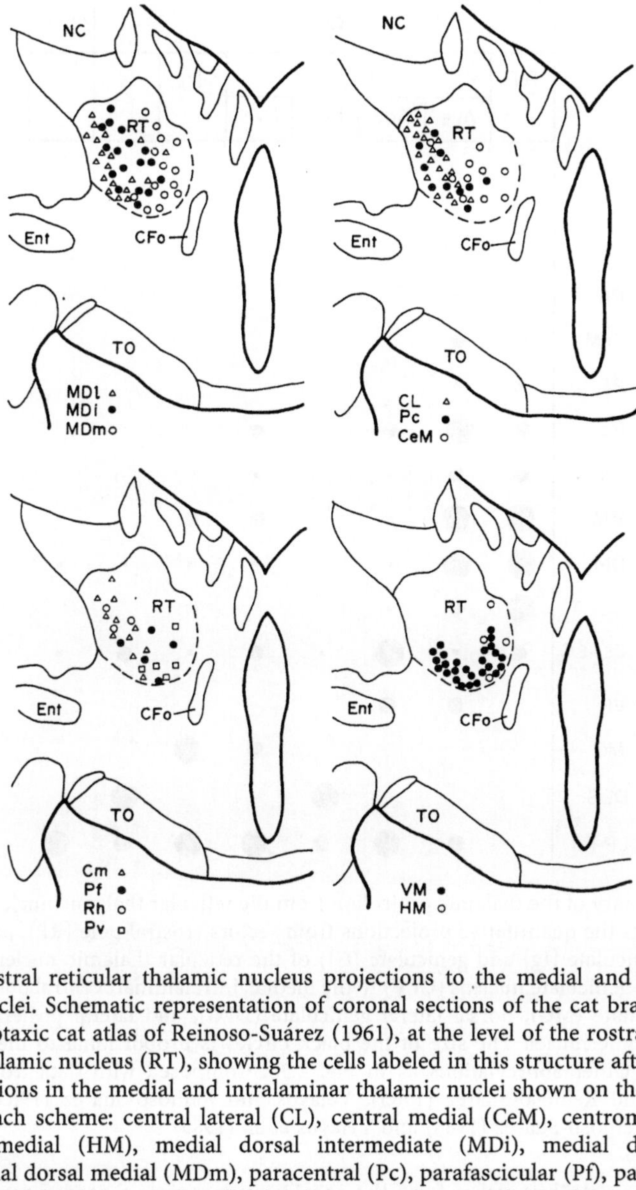

Fig. 2.36 Rostral reticular thalamic nucleus projections to the medial and intralaminar thalamic nuclei. Schematic representation of coronal sections of the cat brain according to the stereotaxic cat atlas of Reinoso-Suárez (1961), at the level of the rostral pole of the reticular thalamic nucleus (RT), showing the cells labeled in this structure after retrograde tracer injections in the medial and intralaminar thalamic nuclei shown on the left inferior corner of each scheme: central lateral (CL), central medial (CeM), centromedian (Cm), habenular medial (HM), medial dorsal intermediate (MDi), medial dorsal lateral (MDl), medial dorsal medial (MDm), paracentral (Pc), parafascicular (Pf), paraventricular (Pv), rhomboidal (Rh), ventral medial (VM). From Velayos et al. (1989)

The long latencies of the recruiting responses elicited in the cat cortex after low frequency stimulation of the intralaminar and related nuclei led to a search for possible routes between these nuclei and the cortex. However, the thalamic projections to superficial layers of these cerebral cortex regions do not arise from the

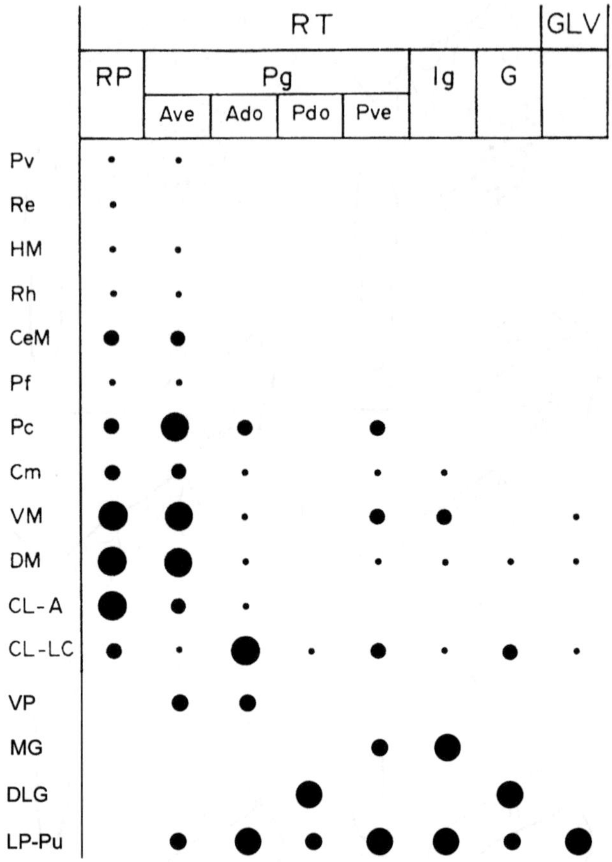

Fig. 2.37 Summary of the thalamic projection from the reticular thalamic nucleus. Schematically presents the quantitative projections from sectors (rostral pole [RP], pregeniculate [Pg], infrageniculate [Ig] and geniculate [G]) of the reticular thalamic nucleus (RT) and ventral lateral geniculate nucleus (GLV) to the medial, intralaminar, ventral posterior (VP), medial geniculate (MG), dorsal lateral geniculate (DLG), and lateral posterior-pulvinar (LP-Pu) thalamic nuclei. The size of the black circles is proportional to the amount of labeled cells. Other abbreviations as in Fig. 2.36. Composed with data from Jiménez-Castellanos and Reinoso-Suárez (1985), Velayos and Reinoso-Suárez (1985), Rodrigo-Angulo and Reinoso-Suárez (1988) and Velayos et al. (1989)

intralaminar nuclei but rather from the paralaminar region, that is, from the ventral medial, ventral anterior, and ventral lateral nuclei; therefore, a large number of subcortical links have been proposed between these thalamic structures, ranging from direct intrathalamic connections to a longer circuit through the basal ganglia (Buchwald et al. 1961; Horvath and Buser 1972). Jiménez-Castellanos and Reinoso-Suárez (1985), Velayos et al. (1989), and Avendaño et al. (1990) proposed a circuit that followed the intralaminar nuclei projections to the reticular thalamic

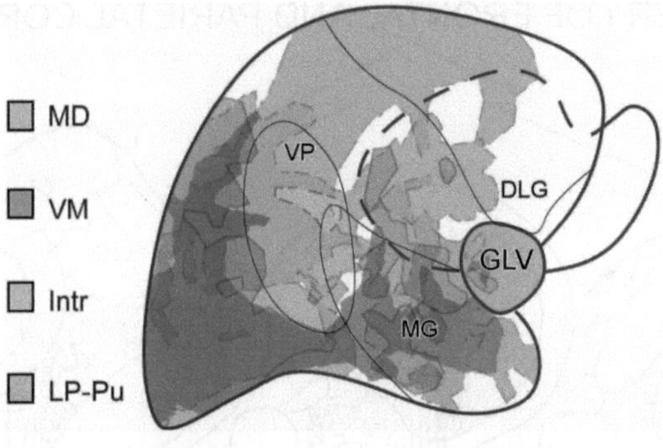

Fig. 2.38 Graphic representation of the reticular thalamic nucleus projection to individual thalamic nuclei. In a schematic lateral representation of the reticular thalamic nucleus, we have projected in different transparent colors the extent of the reticular nucleus projections to the medial dorsal (MD), ventral medial (VM), intralaminar (Intr), and lateral posterior-pulvinar (LP-Pu) thalamic nuclei, according to data from Jiménez-Castellanos and Reinoso-Suárez (1985), Velayos and Reinoso-Suárez (1985), Rodrigo-Angulo and Reinoso-Suárez (1988), and Velayos et al. (1989). *Thin lines* show the areas of projection to the medial geniculate (MG), ventral posterior (VP), and dorsal lateral geniculate nuclei according to data from Jones (1975)

nucleus and the projection from the latter to the paralaminar ventral medial, ventral anterior, and ventral lateral nuclei. The GABAergic neurons of the reticular thalamic nucleus have been shown (see above) to be the pacemaker for spindle oscillations and, through their thalamic projections, would impose their rhythm on the paralaminar M neurons projecting to layer I of the frontal and parietal cortices (Figs. 2.33, 2.36–2.38, and 2.40). This requires, as happens in stage 2 of sleep, the decrease of brainstem activating system action on the reticular thalamic nucleus neurons, which consequently fire at their intrinsic activity level, as well as its decreased action on the paralaminar and other thalamic cells. Taken together these decreases allow the slow oscillatory activity to be expressed in all thalamic neurons. The thalamic neurons, in their turn, project to layer I of the cerebral cortex and impose the slow oscillatory activity on the cortex, which has also been released from the influence of the ascending activating system (Fig. 2.40) (Reinoso-Suárez 1992, 1993).

Stimulation of the cat and monkey cerebellar nuclei produces, with a short latency (about 3 ms), superficial thalamo-cortical responses (surface negative-deep positive potentials) in the frontal and parietal cortices with specific species differences (Sasaki et al. 1972, 1976). However, suppressing these impulses by lesion to the brachium conjunctivum tract produces, together with a bilateral activation of the EEG, the appearance of an intermediate- to high-voltage EEG activity in the theta band (7–13 c/s, bigger of 250 μV) that can be recorded in the cortical areas (mainly the contralateral frontal and parietal cortices) reached by the projections from the thalamic nucleus that receive cerebellar afferents (Figs. 2.13c and 2.39) (Camacho-Evangelista 1962; Camacho-Evangelista and Reinoso-Suárez

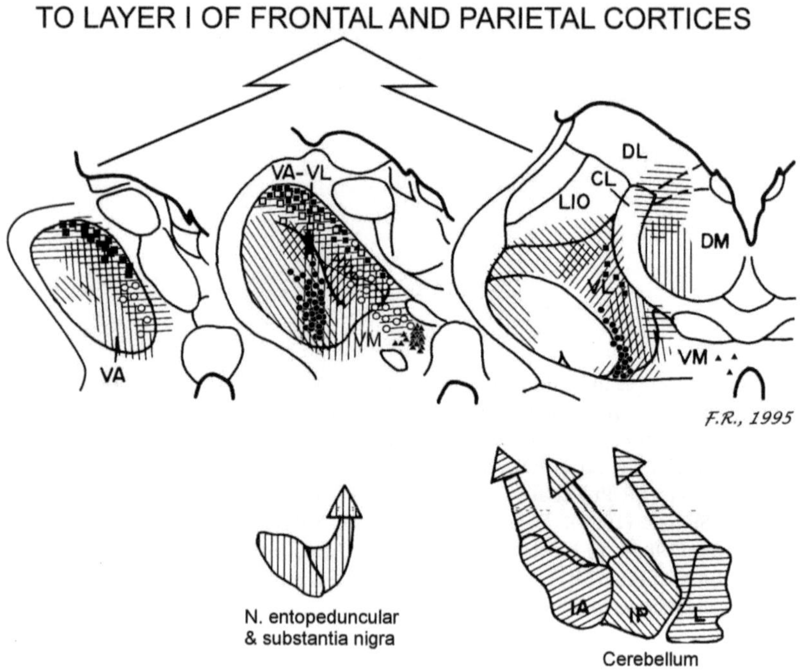

Fig. 2.39 Overlapping at thalamic nuclei of the cerebellar and basal ganglia afferents and the origin of the projections to layer I of the frontal and parietal cortices. The relative locations of the thalamocortical paralaminar neurons projecting to layer I of the cerebral cortex are presented according to the data provided by Oka et al. (1982) (*open circles* are projections to frontal cortex and *open squares* are projections to parietal cortex), Rausell and Avendaño (1985) (*black circles* projections to motor cortex and *black triangles* to prefrontal cortex), and Avendaño et al. 1990 (*black square* to parietal cortex). *CL* central lateral thalamic nucleus; *DL* dorsal lateral thalamic nucleus; *DM* dorsal medial thalamic nucleus; *IA* anterior interposed cerebellar nucleus; *IP* posterior interposed cerebellar nucleus; *L* lateral cerebellar nucleus; *LIO* lateral intermedio oral thalamic nucleus; *VA* ventral anterior thalamic nucleus; *VL* ventral lateral thalamic nucleus; *VM* ventral medial thalamic nucleus

1965; Reinoso-Suárez 1992, 1993). A similar EEG synchronization, recorded most conspicuously in the frontal and parietal cortices, appears particularily ipsilaterally to a lesion of the brachium conjunctivum tract at the level of the red nucleus (Fig. 2.22) (Reinoso-Suárez 1952, 1954). This intermediate- to high-voltage slow frequency activity may also be recorded in the thalamic paralaminar nuclei that receive afferents from the cerebellum; it appears as though suppression of the cerebellar afferents releases these nuclei and allows them to fire at a possibly intrinsic frequency, modulated by thalamic and other extrathalalamic connections (reticular thalamic, cortical, basal forebrain, hypothalamic, brainstem), that project to layer I and impose this bioelectrical activity on the frontal and parietal cortices (Fig. 2.43) (Reinoso-Suárez 1992, 1993).

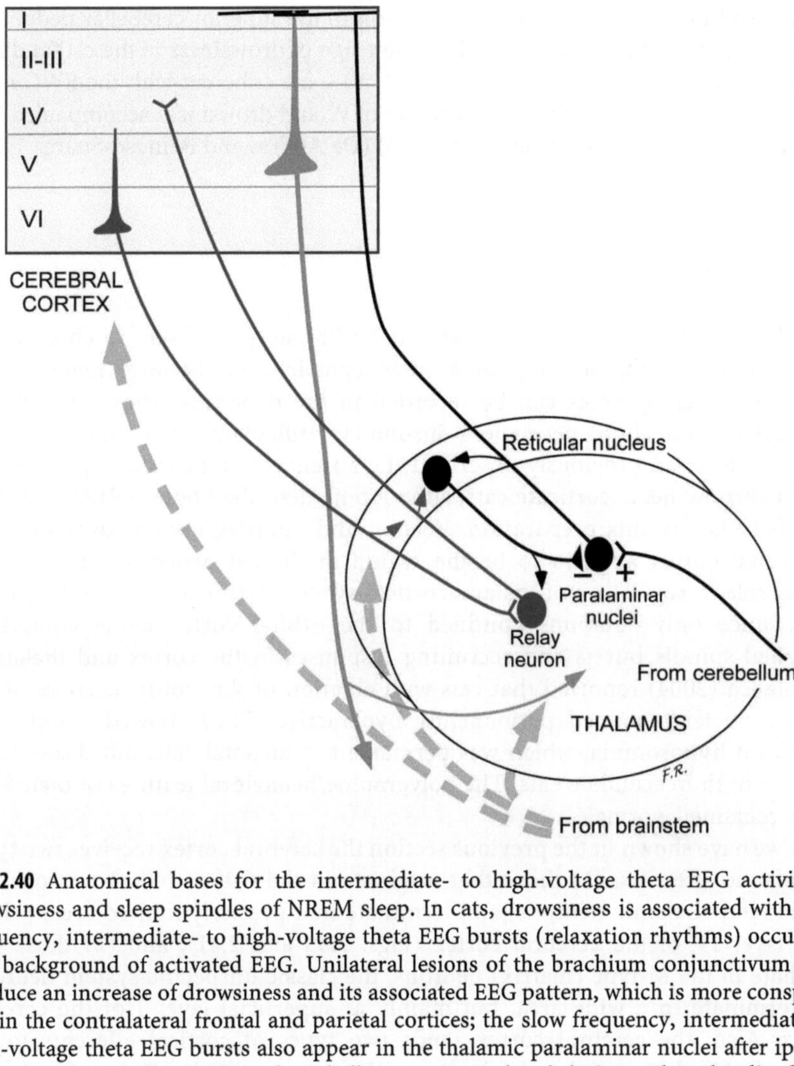

Fig. 2.40 Anatomical bases for the intermediate- to high-voltage theta EEG activity of drowsiness and sleep spindles of NREM sleep. In cats, drowsiness is associated with slow frequency, intermediate- to high-voltage theta EEG bursts (relaxation rhythms) occurring in a background of activated EEG. Unilateral lesions of the brachium conjunctivum tract produce an increase of drowsiness and its associated EEG pattern, which is more conspicuous in the contralateral frontal and parietal cortices; the slow frequency, intermediate- to high-voltage theta EEG bursts also appear in the thalamic paralaminar nuclei after ipsilateral red nucleus and contralateral cerebellar superior peduncle lesions. These localized EEG events may have been caused by the suppression of the cerebellar impulses that liberate the paralaminar nuclei neurons and allow them to fire to a specific frequency, modulated by cortical (layers V and VI) pyramidal neurons and thalamic (reticular and intrinsic) neurons, each influenced by the brainstem activating system which, at the same time, the events are defacilitated by the brachium conjunctivum lesion. These thalamic paralaminar neurons projecting to layer I would impose this bioelectrical activity on the frontal and parietal cortices. In the case of the sleep spindles, the decrease of brainstem activating system activity in stage 2 of sleep allows the reticular thalamic nucleus neurons to impose their intrinsic activity at sleep spindle frequency on the paralaminar neurons, which in their turn are liberated from the influence of the brainstem activating system and project to the layer I, transferring the spindle activity to the frontal and parietal cortices that have also been liberated from the influence of the ascending activating system. Constructed with data from Reinoso-Suárez (1954, 1992, 1993) and Camacho-Evangelista and Reinoso-Suárez (1965)

The two bioelectric phenomena after lesion to the superior cerebellar peduncle – EEG activity in the theta band (one significant sign of drowsiness in the cat) and EEG activation (the most expressive EEG sign of W) – are coherent with the SWC modifications consisting in a significant increase of W and drowsiness accompanied by a decrease in REM and NREM sleep (Fig. 2.15) (De Andrés and Reinoso-Suárez 1979).

2.6 Cerebral Cortex

Villablanca (1972, 2004) affirms that true NREM sleep is absent in chronic cats with ablation of the *telencephalon* (diencephalic cats) because neither delta waves nor sleep spindles can be recorded in the thalamus. These animals are hyperactive and show permanent insomnia (Villablanca and Marcus 1972). Jouvet (1962) had previously described the absence of subcortical synchronization in chronic neodecorticate cats, although he described both NREM and REM sleep behavior in this preparation. Velasco and Lindsley (1965) concluded that the orbital cortex appears to be the region of the cat neocortex that plays a crucial role in regulation of thalamocortical synchronizing and integrating functions, since only ablations confined to the orbital cortex alone completely abolished spindle bursts and recruiting responses in the cortex and thalamus. Villablanca (2004) reported that cats with ablation of the frontal cortices or the caudate nuclei remained permanently hyperactive. They showed a mild, but significant hyposomnia, which was permanent in afrontal cats, but disappeared after a month in acaudate cats. The polygraphic/behavioral features of their SWC states remained normal.

As we have shown in the previous section the cerebral cortex receives two types of thalamic afferents: (1) those that originate in the C-type (core) neurons, the classic relay thalamic neurons, and terminate in a precisely limited area, principally layer IV of the cerebral cortex (Figs. 2.33 and 2.41); and (2) those that originate in the M-type (matrix) neurons, the classic unspecific system neurons, and terminate in a wide area, but mainly at superficial layer I of the cerebral cortex. Also, the cerebral cortex sends two types of cortical efferents to the thalamus: (1) those originating in layer VI neurons carrying a response to the C-type neurons afferents; they are found both in the primary sensory and motor cortices as well as in the association cortices responding to the C-type afferents from both the first-order (relay) and the higher-order (associative) thalamic nuclei, respectively; and (2) those originating in layer V, which are found mainly in the association cortices and project to subcortical nonthalamic structures but also provide collaterals that reach the thalamo-cortical projection neurons of the essentially higher-order thalamic nuclei (Figs. 2.33 and 2.41).

These layer V afferents provide information to the higher-order thalamic nuclei in the same way that the motor or sensory afferents provide information to the thalamocortical neurons of the first-order thalamic nuclei without giving collaterals

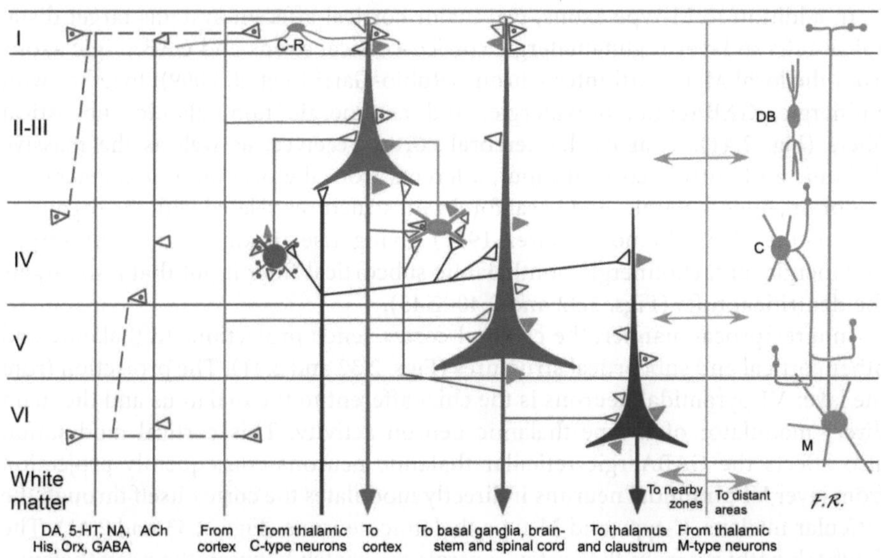

Fig. 2.41 Schematic representation of the fiber and neuronal organization of the cerebral cortex. The projection neurons, pyramidal neurons in layers II–III, V, and VI, are represented in different colors according to their origin and targets. The two types of interneurons are represented in different colors: (1) the excitatory interneuron, a spiny stellate cell, is in pink; and (2) the inhibitory interneuron in light green. We show four specific examples of inhibitory interneuron: two dendrite- and tuft-targetting cells (Cajal-Retzius [C-R] and Martinotti [M] neurons), one dendrite-targetting cell (double bouquet [DB] neuron), and one axon targetting cell (chandelier [C] neuron). Afferent fibers with cortical and subcortical origins are represented in different colors and their specific distributions. The ample distribution of dopaminergic (DA), serotonergic (5-HT), noradrenergic (NA), histaminergic (His), orexinergic (Orx), and GABAergic (GABA) fibers originating in brainstem, diencephalic, and basal prosencephalic structures is represented by their terminals, as is the topographically organized terminals of the basal forebrain cholinergic (Ach) fibers. Thalamocortical fibers targeting cortical layers I (M-type) and IV (C-type) are also represented. I to VI, cortical layers one to six. *Open triangles*, excitatory terminals; *solid triangles*, inhibitory terminals

to the reticular thalamic nucleus, in contrast, other cortico-thalamic afferents do (Fig. 2.33). Thalamic input to apical tufts of layer V, III, and II pyramidal neurons arises in a distinct population of thalamocortical M-type neurons. The axons from large numbers of different M-type neurons converge on layer I in combinations specific for each area, forming a thick subpial mesh studded with bouton-like swellings (Jones 2001; Rubio-Garrido et al. 2009) (Figs. 2.33 and 2.41). The thalamic innervation of cortical layer I is at least as heavy as that of layer IV; every spot of cortical layer I receives the combined input from multiple thalamic nuclei (Rubio-Garrido et al. 2009). All thalamocortical terminals in layer I form asymmetric synapses onto spines, presumably belonging to distal apical dendritic tufts from layers II, III, and V pyramidal neurons (Arbuthnott et al. 1990; Kubota et al. 2007).

In addition to M-type axons, two major cortical afferent systems target distal apical tufts in layer I: glutamatergic cortico-cortical axons and GABAergic axons from the local Martinotti interneurons (Rubio-Garrido et al. 2009), together with aminergic, GABAergic, orexinergic, and cholinergic nonthalamic subcortical fibers (Fig. 2.41). That is, the cerebral cortex receives, as well as the massive thalamic and cortico-cortical input, afferents from the oral pontine tegmentum, midbrain, hypothalamic, and basal forebrain structures related with the organization of the SWC (Reinoso-Suárez 1997) giving rise to aminergic, GABAergic, orexinergic, and cholinergic nonthalamic subcortical fiber input that also targets the dentritical tufts (Figs. 2.32 and 2.40–2.43).

In a reciprocal manner, the cerebral cortex sends projections to thalamic and other cortical and subcortical structures (Figs. 2.32 and 2.41). The projection from the layer VI pyramidal neurons is the chief afferent to the thalamus and the main direct modulator of C-type thalamic neuron activity. This cortical modulation also affects the GABAergic reticular thalamic neurons consequently projection from layer VI piramidal neurons indirectly modulates the cortex itself through the reticular nucleus, C-type and M-type thalamic neurons (Figs. 2.33 and 2.41). The feedback pathway from the cortex controls the switch between the relay thalamic neurons response modes that allow the thalamus to provide a dynamic relay modulating the nature and format of the information reaching the cortex (Sherman 2001). This cortical descending modulation coexists with modulation exerted by the brainstem, hypothalamic, and basal forebrain ascending reticular impulses on the cortical projecting and reticular thalamic neurons (Figs. 2.33, 2.41, and 2.43) (Sherman 2001; Nicolelis and Fanselow 2002).

However, in carnivores and primates most of the thalamus is comprised of the so-called higher order thalamic nuclei and it is the C-type cortical projecting neurons that project to the association areas in the cerebral cortex and receive the feedback pathway from the pyramidal neurons in layer VI of these cerebral cortex areas. These impulses may also control the switch between these relay thalamic neuron response modes that affects the nature and format of information reaching the cortex. However, the information reaching the thalamic projection neurons of these nuclei comes from the layer V pyramidal neurons of the cerebral cortex (Fig. 2.33), some from the areas of each nucleus C-type neuron projection, but mainly from other areas including very distant cortices (Velayos et al. 1993; Rodrigo-Angulo and Reinoso-Suárez 1995). These are the same layer V neurons that receive afferents in their apical tufts from thalamocortical M-type neurons from very different thalamic nuclei (Fig. 2.33) (Rubio-Garrido et al. 2009).

In addition to other functions, these complex thalamocortical relationships may be the foundation for the importance of the thalamus and cerebral cortex in the bioelectrical organization of drowsiness and NREM sleep phases; the thalamic projections to the layer I cerebral cortex and the response from layer V neurons to the thalamus are particularly important. An example is our description above in relation to the slow theta waves of drowsiness and the sleep spindles of stage 2 of sleep (Figs. 2.33, 2.40, and 2.43) (Reinoso-Suárez 1954, 1992, 1993;

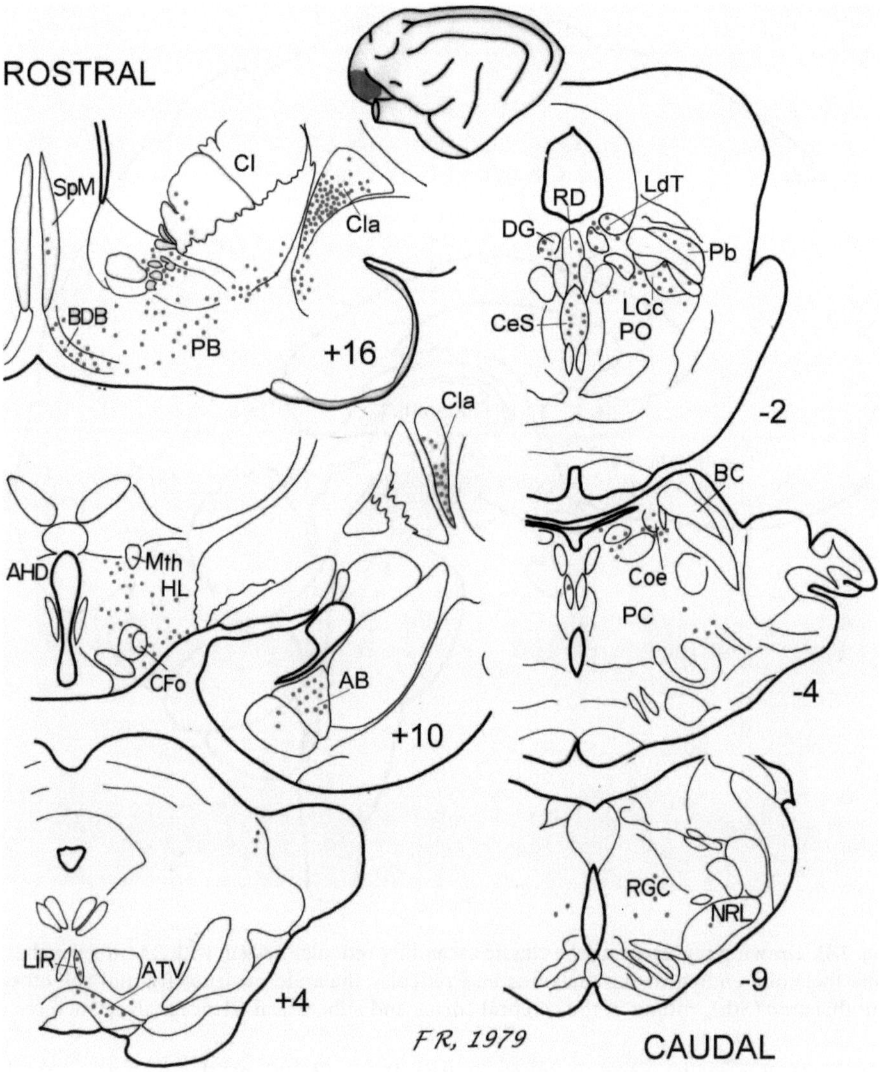

Fig. 2.42 Subcortical nonthalamic projections to the cat prefrontal cortex. Neurons labeled in the subcortical nonthalamic structures after a horseradish peroxidase injection in the cat prefrontal cortex (represented in the diagram of the lateral cortical surface placed at the top of the figure) are schematically presented in coronal sections of the cat brain, ordered from rostral to caudal. *AB* basal amygdaloid nucleus; *AHD* dorsal hypothalamic area; *ATV* ventral tegmental area; *BC* brachium conjunctivum; *BDB* Broca's diagonal band; *CeS* central superior nucleus; *CFo* fornix; *Cla* claustrum; *Coe* locus coeruleus nucleus; *DG* dorsal tegmental nucleus; *HL* lateral hypothalamic area; *IC* internal capsule; *LdT* laterodorsal tegmental nucleus; *LiR* rostral linear nucleus; *Mth* mammillothalamic tract; *NRL* lateral reticular nucleus; *PB* basal forebrain; *Pb* parabrachial nucleus; *PC* caudal pontine reticular nucleus; *PO* oral pontine reticular nucleus; *RD* dorsal raphe nucleus; *RGC* gigantocellular reticular nucleus; *SpM* medial septal nucleus

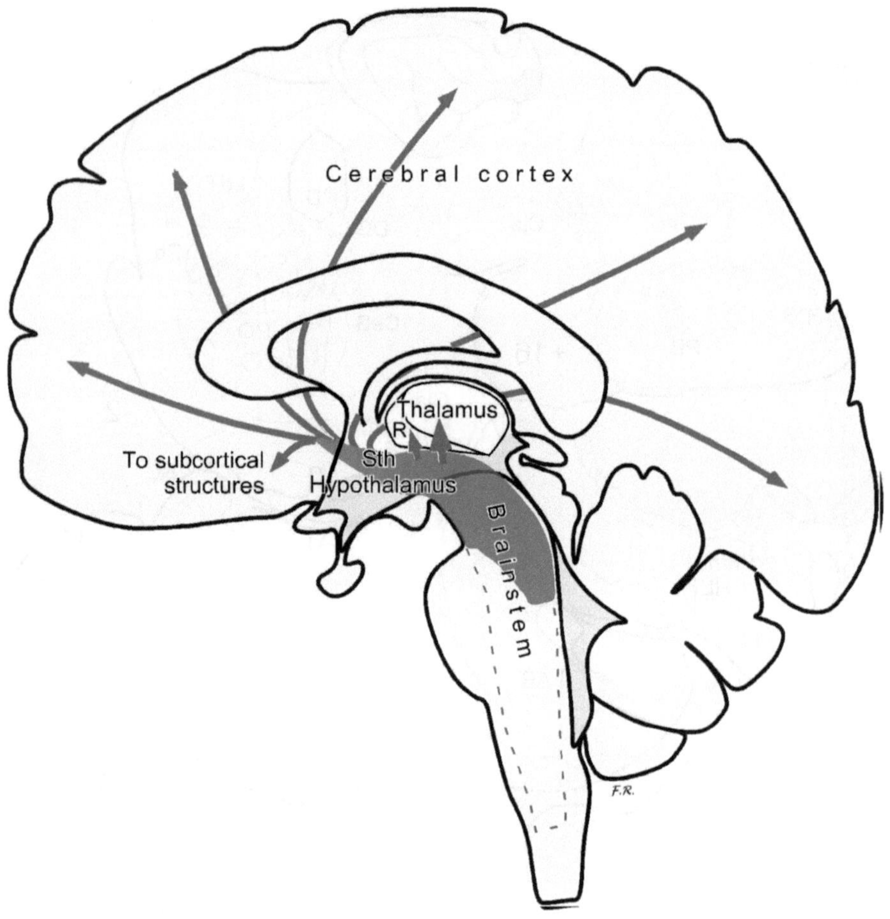

Fig. 2.43 Drawing representing the classic ascending reticular system with its two branches, one thalamic, ending in the thalamus and reticular thalamic nucleus (R), and the other subthalamic (Sth), ending in the cerebral cortex and subcortical telencephalic structures

Camacho-Evangelista and Reinoso-Suárez 1965; Sasaki et al. 1972; Oka et al. 1982; Avendaño et al. 1990).

2.7 Final Commentary

To our understanding the thalamus and the cerebral cortex, closely and abundantly interconnected, form a nondivisible unit in brain function (Figs. 2.32 and 2.33). This is why most brainstem, hypothalamic, and basal forebrain nonspecific sensory or motor systems act directly on both the thalamus and the cortex; these regions are the targets of the two branches of the classic ascending reticular system: one thalamic, ending in the thalamus, another subthalamic, ending in the cerebral cortex and subcortical telencephalic structures (Figs. 2.32, 2.33, 2.41, and 2.43). Both branches influence the thalamocortical relationship and consequently the final response of this thalamocortical system and its modulation of the basal forebrain, hypothalamic, and brainstem structures related with the organization of the SWC.

It is important to emphasize that the two components (thalamus and cerebral cortex) of this top level in the organization of the nervous system are necessary for the expression of the most significant bioelectric (sleep spindles and slow wave EEG) and behavioral signs that define NREM sleep. Without either of these two structures true NREM sleep is absent. However, the basal forebrain-anterior hypothalamic region appears to be essential for the induction of the NREM sleep state; this region is closely connected with the thalamus and cerebral cortex (Fig. 2.32). Other regions situated in the caudal pontine and in the medulla oblongata tegmenti also participate in the organization of NREM sleep; these structures receive connections from the cerebral cortex and project to more distant rostral brainstem structures and the thalamus.

The oral pontine tegmentum is essential for the organization of REM sleep, requiring the contribution of brainstem structures located more caudally. The oral pontine tegmentum is bidirectionally connected with the caudal brainstem structures and spinal cord, the cerebellum, the midbrain, hypothalamus, thalamus, and basal forebrain (Fig. 2.32).

The waking structures are situated between the rostral and caudal NREM sleep inducing structures, that is to say, in the oral pontine and midbrain tegmenti and posterior lateral hypothalamus. They receive and send connections from and to the rostral and caudal NREM and REM sleep inducing areas (Fig. 2.32).

Other nervous system structures such as the sensory nerves, the spinal cord, the cerebellum, the amygdala, and the basal ganglia may also modulate the functional patterns of the different phases of the SWC (Fig. 2.32).

The classic studies of lesion, stimulation, and anatomical connectivity provide a wide vision of the brain regions responsible for the organization of the SWC and the relationships among its different states. However, the matter is even more complex: every one of these regions includes many structures that presumably have different functions, connections, and neurochemical properties; also, the same structure may contain neurons participating in the organization of different phases of the SWC and may even use different neurotransmitters. But, as Nicholson (2007) affirms, the SWC as a whole is dependent on the cumulative effect of many systems and it is necessary to investigate the interaction of the different brain structures and types of neurons that participate in these interactions to elucidate their true contribution to sleep and W. Therefore, supported by the descriptions made in this chapter, we will present in the next chapter a detailed review and update of the structures involved in the W phase of the SWC, as well as of their morphological, functional and chemical characteristics, and their connections.

Chapter 3
Functional Anatomy of Wakefulness

The wakefulness (W) phase of the SWC in humans usually happens during the day. Only during W are we aware of ourselves and our environment; that is, only during W do we feel, think, and work in full knowledge of what we do and fully use our senses and faculties. During W is when we properly process all the information that reaches the thalamus and the cerebral cortex from the sensory organs and from the propioceptors and intraceptors distributed in our organism. W is therefore necessary for our nervous system to support the organization of cognitive processes in their different categories. The state of W primes the nervous system to perform these functions.

Surprisingly, as we have seen in Chap. 2, the structures responsible for the organization of the W phase of the SWC are located in the oral pontine and midbrain tegmenti and lateral posterior hypothalamus. As we described, the active W phase depends: (1) on influences descending to the lower brainstem and spinal cord that organize behavioral arousal with adequate muscle tone; and (2) on influences ascending from the brainstem to the thalamus and cerebral cortex that stimulate forebrain activation. This chapter will describe the different components of this brainstem–hypothalamic coordination entity: its neurons; the neurotransmitters it uses; its connections and the mechanisms through which it attains forebrain arousal. Finally, we will describe the structures located far from this rostral brainstem–hypothalamic region that participate in specific aspects of W.

As noted in the first chapter, in man W shows low-voltage and high-frequency or intermediate-frequency (13–30 Hz) EEG activity, sometimes referred to as activated or desynchronized activity; in addition, there is a rhythmic activity of 30–60 Hz (gamma activity) that appears in all areas of the cerebral cortex during vivid attention or motor processing. Robust muscle activity is usually present in the EMG. Eye movements and frequent movement artifacts are present in the recording. In the cat, our most frequent experimental animal, W shows an activated electrocorticogram (ECoG) with almost continuous low-voltage fast activity, theta activity in the hippocampus, tonic EMG, numerous eye movements in the EOG, and absence of PGOs.

3.1 The Brainstem–Hypothalamic Wakefulness Structures and Their Neurotransmitters

As already mentioned, Bremer (1937) was the first to attribute W to tonic impulses ascending through the brainstem that maintained forebrain W. He assumed that such impulses were carried through the classical sensory pathways that ascend within the brainstem; he proposed that these ascending impulses were temporarily disrupted during sleep and permanently disrupted in coma, in both cases with a synchronized electroencephalogram (Fig. 2.3). In 1949, Moruzzi and Magoun showed that those tonic impulses that awaken the brain and maintain the state of W originated from the reticular formation of the brainstem (Fig. 2.5). They described the "ascending reticular activating system" as the structure responsible for the characteristic cortical activation to the W state through a diffuse and nonspecific projection to the cerebral cortex; this system contrasts with the very specific cortical projection of the sensory pathways that ends in the thalamic relay nuclei specific to each particular sensory modality. These specific sensory pathways influence the brainstem reticular formation through collaterals that produce tonic impulses important for the arousal reaction and maintenance of W. In fact, all the numerous ascending and descending pathways traveling along the brainstem contribute to the activity of the reticular formation, and those pathways originating in the own cerebral cortex itself assume a particularly major role.

We should not forget that some waking influences can ascend from the medulla oblongata and the very caudal pontine reticular formation, since, as described in the previous chapter, chemical and electrical stimulation of these structures increase W (Moruzzi and Magoun 1949; Baghdoyan et al. 1984; Garzón 1996); also, the preparation caudal to midpontine transections may show a brief period of primitive arousal immediately after the lesion and later periodic behavioral activation, possibly resembling a very elemental waking state (Fig. 2.1) (Siegel et al. 1986; Villablanca 2004). However, after the descriptions of Batini et al. (1959) and Camacho-Evangelista and Reinoso-Suárez (1964), it was determined that the most important tonic brainstem waking influence, the one that was necessary to maintain the cat awake, originated caudal to the midbrain in the oral pontine tegmentum rostral to the midpontine pretrigeminal plane, i.e., in the oral pontine tegmentum (Figs. 2.1 and 2.7). Later, it was demonstrated that the ventral part of the oral pontine reticular nucleus is related with the organization of REM sleep (Reinoso-Suárez et al. 1994, 2001); consequently, the main origin of the brainstem waking impulses that would ascend to the forebrain but that are suppressed in the *cerveau isolé* preparation are the structures located in the dorsal oral pontine and caudal midbrain dorsal tegmenti, that is, in the dorsal mesopontine tegmentum. However, the midbrain reticular formation and, above all, the posterior-lateral hypothalamus play a critical role in the maintenance of W, since the posterior-lateral hypothalamus is the main area that is responsible for the waking periods in the chronic *ceveau isolé* preparation (see Chap. 2).

3.1.1
The Dorsal Mesopontine Tegmentum

The dorsal mesopontine tegmentum, principal origin of the ascending brainstem arousal system, is a biochemically heterogeneous region containing noradrenergic, serotonergic, and cholinergic neurons. Traditionally, these neurons are considered to be located in specific groups with known preferential, but frequently mixed, neurotransmitters. Among these neuronal groups, most authors distinguish the dorsal (DR) and central superior (CS) raphe nuclei, which contain serotonergic neurons; the locus coeruleus (LC), rich in noradrenergic neurons; and the pedunculopontine (PpT) and laterodorsal (LdT) tegmental nuclei, containing cholinergic neurons (Reinoso-Suárez 1997). Nevertheless, and along with many of other types of neurons, most of these nuclei contain a great variety of neuropeptide-containing neurons, and an appreciable amount of GABAergic and glutamatergic neurons (Sutin and Jacobowitz 1988; Clements and Grant 1990; Jones 1991; Sakai 1991; Lavoie and Parent 1994; Nitz and Siegel 1997; Maloney et al. 1999; Gervasoni et al. 2000; Brown et al. 2008).

All types of mesopontine tegmentum neurons are widely mixed in the different nuclei; thus in the cat and only referring to cholinergic, adrenergic, and serotonergic neurons, there are: the locus coeruleus complex (LCC) comprised of a diverse nuclei, most of which, in addition to noradrenergic neurons, also contain serotonergic neurons [locus coeruleus α (LCα), perilocus coeruleus α (Pα), and subcoeruleus (SCoe)] and cholinergic neurons (LCα and Pα). Cholinergic, serotonergic, and noradrenergic neurons are also observed among the fibers of the central tegmental fasciculus, a structure that at this level is considered by some authors to be part of the LCC. The LdT nucleus contains only a limited number of the dorsal mesopontine tegmentum cholinergic neurons, but it also contains an appreciable amount of serotonergic and some noradrenergic neurons (Fig. 3.1) (see Reinoso-Suárez et al. 2001 and Rodrigo-Angulo et al. 2000, 2005). The parabrachial (Pb) nuclei contain abundant noradrenergic, cholinergic, and serotonergic neurons, but possibly also neurons with dopamine and other neurotransmitters. Serotonergic neurons are also found in the periaqueductal gray matter, the tegmental dorsal nucleus, and the oral pontine reticular nucleus. The DR and CS and every neuron located near these nuclei form part of the B8 serotonergic cell group of Dahlström and Fuxe (1964), whereas the more laterally located serotonergic neurons like those of the oral pontine reticular nucleus are part of the B9 serotonergic group (Fig. 3.1). The abundant serotonergic neurons located outside the raphe nuclei are the main origin of the serotonergic projections to other brainstem structures, whereas the serotonergic DR and CS neurons appear to be the origin of the serotonergic projections to midbrain, hypothalamus, thalamus,

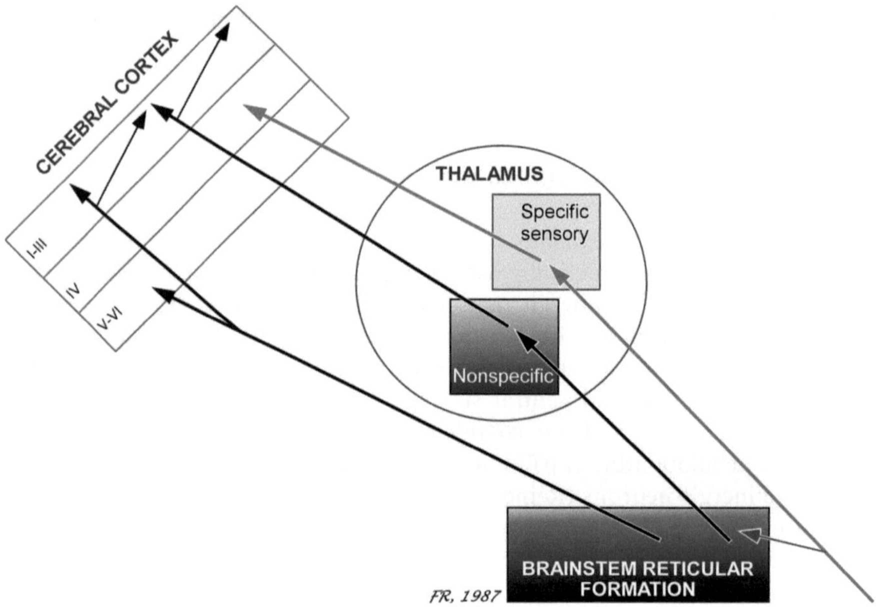

Fig. 3.1 Schematic drawing of the classic "ascending reticular activating system". The tonic impulses that awaken the brain and maintain the state of W originate in the reticular formation of the brainstem. The thalamus and the cerebral cortex are the final targets of the two branches of the reticular ascending activating system: the thalamus in the dorsal or thalamic branch and the cerebral cortex in the ventral or subthalamic branch. The thalamus in its turn projects to the cerebral cortex through a diffuse and nonspecific projection system; this system contrasts with the specific cortical projection of the sensory pathways that make a station in the thalamic relay nuclei specific to each particular sensory speciality and then project on layer IV of a specific cortical area. The specific sensory pathways influence the brainstem reticular formation through collaterals that contribute with tonic impulses to awakening and W. Modified from Reinoso-Suárez (1971)

and cerebral archi- and neo-cortex (Figs. 2.18–2.21, 2.35, and 2.42) (Reinoso-Suárez and Llamas 1975; Reinoso-Suárez 1997; Rodrigo-Angulo et al. 2000).

Traditionally, LdT and PpT were considered the chief origin of cholinergic projections to other ipsilateral brainstem and prosencephalic structures; nowadays, it is known that the major sources of cholinergic projections to the ventral oral pontine tegmentum in the cat are the ipsilateral and contralateral PpT and LCC, and this is supported by the number of cholinergic neurons in each of these nuclei (Rodrigo-Angulo et al. 2005). The origin of the dorsal mesopontine tegmentum cholinergic projections to midbrain, thalamus, hypothalamus, and basal forebrain structures would follow similar distribution. Therefore, it is perhaps more appropriate, as Mesulam (1995) does in the monkey, to call the cholinergic neurons associated with the PpT the CH5 group and those associated with the LdT nucleus the CH6 group, since cholinergic neurons occupy a wide area that greatly exceeds the borders of either of these nuclei (Fig. 3.1).

As we have described in the previous chapter, the *noradrenergic* neurons of the dorsal mesopontine tegmentum, most of which constitute the A6 neuronal group of Dahlström and Fuxe (1964), project to widespread areas of the forebrain (hypothalamus, thalamus, basal forebrain, basal nuclei, hippocampus, and cerebral cortex), brainstem, cerebellum, and spinal cord, where their axons reach the sensory relay, somatic, and visceral motor neurons. Jones (2008) reports that noradrenaline has either excitatory or inhibitory actions upon postsynaptic neurons depending upon the receptors that are activated; thus, it can directly excite the thalamo–cortical relay neurons and the cholinergic basal forebrain neurons, as well as cortical pyramidal neurons, and also brainstem and spinal motor neurons through α1-receptors. Noradrenaline inhibits sleep-promoting neurons through α2-receptors. Thus, when dorsal pontine tegmentum noradrenergic neurons are excited, they can simultaneously stimulate cortical activation and behavioral arousal with adequate muscle tone.

The activity of these three types of neurons, cholinergic, serotonergic, and noradrenergic, probably plays, together with other brain functions, an outstanding role in cortical desynchronization and behavioral W as part of the "ascending reticular activating system" but, as we will describe below, the arousal system is very complex and needs many other components to maintain normal W.

Acetylcholine. It has long been known that enhancement of acetylcholine levels, by inhibition of the catabolic enzyme acetyl-cholinesterase, evokes waking with cortical activation (Karczmar et al. 1970). It is also known that the dorsal mesopontine tegmentum cholinergic neurons fire very quickly during W and REM sleep and are much less active during NREM sleep (el Mansari et al. 1989); in contrast, the mesopontine aminergic neurons, both serotonergic and noradrenergic, fire fastest during W, slow down during NREM sleep and stop altogether during REM sleep (McGinty and Harper 1976; Aston-Jones and Bloom 1981; Aston-Jones et al. 1991; for a revision see McCormick 1992). However, while this is true in general terms, we now know that there are other types of dorsal mesopontine neurons, for example, cholinergic neurons that behave differently in the various phases of the SWC. Using single-unit recording in unanesthetized cats and rats, three different dorsal mesopontine tegmentum cholinergic neuron firing patterns have been identified in the SWC (Nelson et al. 1983; El Mansari et al. 1989; Kayama et al. 1992). These patterns are expressed by: (1) a small group (5–9%) of mesopontine cholinergic neurons that show phasic activities in relation to PGO waves, with phasic field potentials recorded during REM sleep (PGO-on neurons) (Steriade et al. 1990); (2) a larger group of cholinergic neurons (more than 40%), which show higher firing rates in REM sleep than in W and NREM sleep (REM-on neurons); and (3) the largest group of mesopontine cholinergic neurons (more than 50%), which show higher firing rates in W and REM sleep than in NREM sleep (W/REM-on neurons) with most of the latter exhibiting the highest firing rates in W. In addition, it is important to emphasize that most mesopontine neurons may utilize more than one neurotransmitter; for example, in cholinergic neurons, a great variety of neurotransmitters colocalize with acetylcholine such as glutamate, atriopeptin, CRF, substance P, or nitric oxide, etc. (Lavoie and Parent 1994; Semba 1999). Glutamate appears to be the main excitatory neurotransmitter of many dorsal mesopontine tegmentum cholinergic neurons and this is also supposed to be true in the basal forebrain and oral pontine reticular formation (Rasmusson et al. 1996; Semba 1999; Fournier et al. 2004). In spite of this complex organization, we know that the mesopontine cholinergic W/REM-on neurons participate significantly in behavioral W and EEG activation through their projections to midbrain, thalamus, hypothalamus, and basal forebrain and very scarce direct projections to the cerebral cortex (Fig. 2.42) (Foote and Morrison 1987; Reinoso-Suárez 1997).

Serotonin and Noradrenaline. Kayama and Koyama (2003) have investigated the specific forebrain actions of the noradrenergic projection originating in the locus coeruleus, the serotonergic projection from the DR nucleus, and the cholinergic projection from neurons in the dorsolateral mesopontine tegmentum by observing the effects of stimulating these brainstem nuclei. They conclude that the projection from the locus coeruleus is an arousal system, since the noradrenergic neurons are active specifically during waking, and activation of the noradrenergic projection excites upper brain structures. These authors conclude that the functions of the ascending serotonergic projection are mysterious, since their action on the upper brain is inhibitory in spite of the waking-specific activity of these neurons. In relation to the cholinergic mesopontine group, they affirm that although a group of cholinergic neurons may participate in the induction of REM sleep, the cholinergic projection to the forebrain may also have another role, consisting in the induction of a rapid, transient elevation of the vigilance level through a phasic response to novel, unfamiliar stimuli.

There are very abundant and heterogeneous afferent connections to the different types of neurons composing the dorsal mesopontine tegmentum. There are connections from the cerebral cortex, the basal forebrain, the hypothalamus, cerebellum, brainstem, and spinal cord, and all of them are involved in the modulation of neuronal activity by this region. The rich intrinsic connections among dorsal mesopontine tegmentum cell groups support the close interactions among serotonergic, noradrenergic, and cholinergic neurons of the dorsal mesopontine tegmentum (Koyama and Kayama 1993). It is coherent that, as shown by electrophysiological data, some excitatory substances like carbachol and hypocretins/orexins seem to act on all dorsal mesopontine tegmentum cell types and, despite their biochemically different phenotypes, generally generate excitatory responses (Brown et al. 2006). But other neurotransmitters of arousal-related systems, which belong to neurons that are intrinsic or extrinsic to the region, such as noradrenaline or histamine may also increase the firing rate of other neuron types, such as the serotonergic neurons (Brown et al. 2002). In any case, we can summarize the existing literature by saying that excitation of the dorsal mesopontine tegmentum neurons has three effects (Fig. 3.2): (1) W enhancement

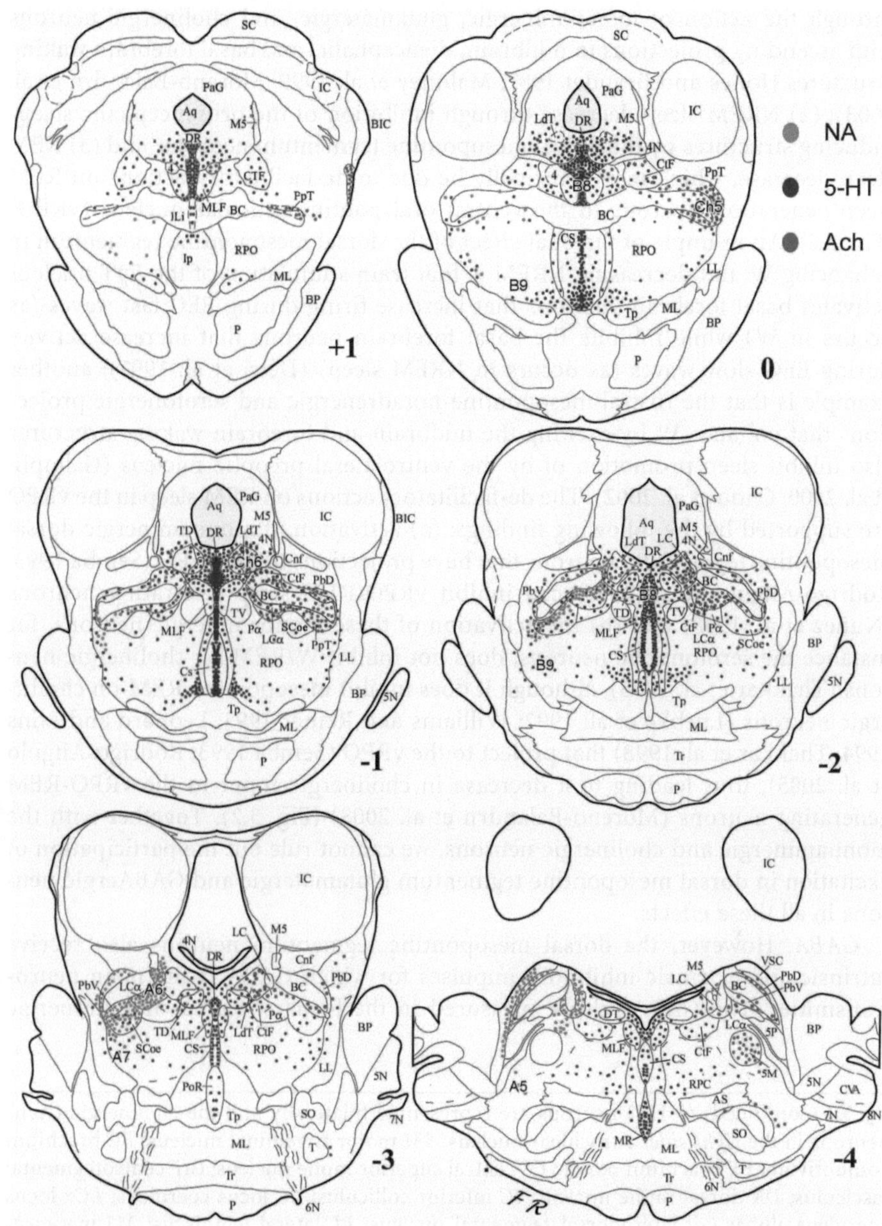

Fig. 3.2 Schematic drawing showing the distribution of the noradrenergic, serotonergic, and cholinergic neurons in the mesopontine tegmentum. Coronal sections of the mesopontine tegmentum from the stereotaxic cat atlas of Reinoso-Suárez (1961), from rostral plane +1 to caudal plane −4, with the widespread and intermingled locations of the cholinergic, noradrenergic, and serotonergic neurons in the different nuclei of this region. The noradrenergic (NA) neurons are represented in the *left side* of each section, the serotonergic

through the action of monoaminergic, glutamatergic, and cholinergic neurons with ascending projections to midbrain, diencephalic, and basal forebrain waking structures (Jones and Beaudet 1987; Maloney et al. 1999; Moreno-Balandrn et al. 2008); (2) NREM sleep decrease through inhibition of the prosencephalic sleep-inducing structures by the dorsal mesopontine tegmentum neurons; and (3) REM sleep decrease, which may principally be due to de-facilitatory actions on REM sleep generation neurons in the ventral oral pontine reticular nucleus (vRPO) (Fig. 3.2). An example of the dual effect of the dorsal mesopontine tegmentum in enhancing W and decreasing NREM is that train stimulation of the PpT nucleus activates basal forebrain neurons that increase firing during EEG fast waves (as occurs in W) while inhibits the basal forebrain neurons that increase activity during EEG slow waves (as occurs in NREM sleep) (Détri et al. 1997); another example is that the rostral mesopontine noradrenergic and serotonergic projections that enhance W by exciting the midbrain and forebrain waking structures also inhibit sleep promotion of by the ventrolateral preoptic nucleus (Gallopin et al. 2000; Chou et al. 2002). The de-facilitatory actions on REM sleep in the vRPO are supported by the following findings: (a) activation of monoaminergic dorsal mesopontine tegmentum neurons that have projections to the vRPO (Semba 1993; Rodrigo-Angulo et al. 2000) can inhibit vRPO-REM sleep generating neurons (Nuñez et al. 1998) and (b) the activation of these monoaminergic neurons, for instance the serotonergic neurons, does not inhibit W/REM-on cholinergic neurons (Thakkar et al. 1998), although it does inhibit mesopontine REM-on cholinergic neurons (Luebke et al. 1992; Williams and Reiner 1993; Leonard and Llins 1994; Thakkar et al. 1998) that project to the vRPO (Semba 1993; Rodrigo-Angulo et al. 2005), thus leading to a decrease in cholinergic input to the vRPO-REM generating neurons (Moreno-Balandrn et al. 2008) (Fig. 3.2). Together with the monoaminergic and cholinergic neurons, we cannot rule out the participation of excitation in dorsal mesopontine tegmentum glutamatergic and GABAergic neurons in all these effects.

GABA. However, the dorsal mesopontine tegmentum neurons also receive intrinsic and extrinsic inhibitory impulses for which GABA is the main neurotransmitter. The GABA release measured in the locus coeruleus noradrenergic

Fig. 3.2 (continued) (5-HT) neurons are represented bilaterally and the cholinergic (Ach) neurons in the *right side*. *4* trochlear nucleus; *5M* motor trigeminal nucleus; *BC* brachium conjuntivum; *BP* brachium pontis; *CS* central superior raphe nucleus; *CtF* centrotegmental fasciculus; *DR* dorsal raphe nucleus; *IC* inferior colliculus; *LC* locus coeruleus; *LCα* locus coeruleus alpha; *Ldt* laterodorsal tegmental nucleus; *LL* lateral lemniscus; *M5* mesencephalic trigeminal tract; *ML* medial lemniscus; *MLF* medial longitudinal fasciculus; *MR* magnus raphe nucleus; *P* pyramidal tract; *Pα* perilocus coeruleus alpha; *PaG* periacueductal gray; *PbD* parabrachial dorsal nucleus; *PbV* parabracial ventral nucleus; *PoR* pontis raphe nucleus; *PpT* pedunculopontine tegmental nucleus; *RPC* caudal pontine reticular nucleus; *RPO* oral pontine reticular nucleus; *SCoe* locus subcoeruleus nucleus; *SO* superior olivary nucleus; *TD* dorsal tegmental nucleus;*Tp* tegmental pontine nucleus; *TR* trapezoid body neurons during the sleep–wakefulness state in freely behaving cats shows higher GABA release during REM sleep than during W while the release levels during NREM sleep fell between levels released during W and REM sleep (Nitz and Siegel 1997). Gervasoni et al. (1998) also demonstrated that noradrenergic neurons in the rat locus coeruleus are tonically inhibited by local GABAergic neurons during sleep. Additionally, it has been speculated that the increase of the GABAergic inhibitory tone present during W may be responsible for decreased activity in DR serotonergic cells during NREM and REM sleep, since iontophoretic application of the $GABA_A$ antagonist bicuculline on DR serotonergic neurons of anesthetized rats induces a tonic discharge during NREM and REM sleep and an increase of discharge rate during quiet waking (Gervasoni et al. 2000). These authors suppose that among the numerous GABAergic afferents to the DR nucleus, the GABAergic neurons located in the lateral preoptic area and the pontine ventral periaqueductal gray including the DR nucleus itself could be responsible for the respective reductions of serotonergic neuron activity in the DR nucleus during NREM and REM sleep. For Kumar et al. (2008), the activation of the hypnogenic median preoptic nucleus neurons contributes to the suppression of waking promoting hypothalamic and dorsal mesopontine tegmentum systems, including the serotonergic neurons in the DR nucleus. On the other hand, the sleep-promoting ventrolateral preoptic nucleus inhibits noradrenergic LC and serotonergic DR neurons through GABAergic and galaninergic connections (Lu et al. 2002; Saper et al. 2005). It has also been proposed that the adrenergic and serotonergic dorsal mesopontine tegmentum neurons would be inhibited by GABAergic neurons located in the sleep-promoting lateral preoptic area during NREM sleep, and that at the onset and during REM sleep these monoaminergic neurons would be inhibited by GABAergic neurons located in the extended ventrolateral preoptic nucleus, the periaqueductal gray, or in the gigantocellular nucleus (Luppi et al. 1998; Saper et al. 2005).

Other neurotransmitters like hypothalamic hypocretins/orexins (Hcrt/Orx) may have regionally selective effects on serotonin release, implying an interaction between Hcrt/Orx and serotonin in the regulation of activities including sleep–wakefulness (Tao et al. 2006). Based on their observations, Lee et al. (2005a) suggested that various hypothalamic neurons differentially project to each subdivision of the DR, a portion of which is Hcrt/Orx-immunoreactive. These Hcrt/Orx-immunoreactive DR-projecting hypothalamic neurons might have wake-related influences over a variety of brain functions involving DR.

However, not only Hcrt/Orx, but also hypothalamic histamine increases the activity of DR serotonin neurons in vivo and in vitro (Brown et al. 2002). Consequently, histaminergic descending projections to the dorsal mesopontine tegmentum promote cortical EEG desynchronization and W, and this was proved because the microinjection of histamine, or a receptor agonist, in the dorsal mesopontine tegmentum increased W, while the injection of a receptor antagonist increased NREM sleep (Lin et al. 1996). Also, the neurons of the oral pontine reticular formation, as a REM sleep generation site, may modulate the activity of mesopontine tegmentum cholinergic neurons (Keifer et al. 1996) and other types of wakefulness-promoting neurons (Fig. 3.2). The anatomical support for this modulation is the rich serotonergic projection that together with a good number of putatively glutamatergic or GABAergic neurons, projects from the vRPO to the site of the dorsal mesopontine tegmentum where a significant number of cholinergic and serotonergic neurons reciprocally send connections back to the vRPO (Rodrigo-Angulo et al. 2000, 2005).

In summary: The activity of the neurons of the dorsal mesopontine tegmentum, main origin of the brainstem ascending arousal system, plays an outstanding role in cortical desynchronization and behavioral arousal. These neurons project rostrally to the midbrain and posterior hypothalamus and project further through two branches: one is thalamic and modulates thalamic nuclei activity, including that of the reticular thlalamic nucleus; and the other is subthalamic. The latter ends in the anterior hypothalamus, basal forebrain, cerebral cortex, and subcortical telencephalic structures (Figs. 2.43 and 3.2). The rostral dorsal mesopontine projections excite midbrain and forebrain structures that participate in achieving proper W and inhibit forebrain structures that are involved in the generation of NREM sleep. The dorsal mesopontine neurons also modulate activity in brainstem structures to inhibit REM sleep generating neurons. In turn, the midbrain and forebrain waking groups excite the mesopontine waking neurons, while the brainstem and forebrain sleep-promoting structures modulate the dorsal mesopontine waking neurons through inhibitory connections (Fig. 3.2).

3.1.2

The Midbrain Tegmentum

The three groups of dorsal mesopontine tegmentum neurons (cholinergic, serotonergic, and noradrenergic) project rostrally to the mid- and forebrain in the company of glutamatergic and GABAergic axons (Reinoso-Suárez 1977, 2005; Saper et al. 2005). In midbrain, they send connections to the superior colliculus, the reticular formation, the dorsal and ventral dopaminergic, and GABAergic cellular groups.

Chapter 2 discussed midbrain reticular formation participation in W organization. Steriade (1981) showed that the midbrain reticular formation neurons projecting to the forebrain have a high tonic discharge rate during cortical activation and decrease their firing rate before the onset of cortical slow wave activity. The forward projection of these neurons excites the thalamocortical projection neurons in the waking period, also exciting the active waking neurons in the posterior lateral hypothalamus and basal forebrain, which are located in the ventral subthalamic branch of the classic "ascending reticular activating system" (Findlay and Hayward 1969; Steriade 1981; Szymusiak and McGinty 1989). The main neurotransmitter for the midbrain reticular formation is glutamate and its neurons should be respectively modulated by inhibitory and excitatory impulses originating in the sleep- and waking-promoting cerebral structures.

The classic experiments with lesions in the substantia nigra (SN) (which contains dopaminergic and GABAergic neurons) and the dopaminergic ventral tegmental area (VTA) have different results in cats and rats. Cats with large diatermocoagulation lesions in the ventral mesencephalic dopaminergic groups showed major decreases in prosencephalon dopamine levels and almost permanent comatose behavior (Jones et al. 1969). However, despite their comatose state, the EEG continued to present alternating synchronized and desynchronized activity. Contrary to this, the destruction of the substantia nigra in rats was followed by an increase in motility (Jones et al. 1969).

Dopamine. Most recent research on dopamine and SWC employs rats as the research animal. Lu et al. (2006) showed that, whereas evidence suggests that dopamine plays an important role in arousal, the firing of the rat dopaminergic neurons in the ventral tegmental area that projects to the prefrontal cortex and striatum does not correlate with overall levels of behavioral W. These authors identified wake-active dopaminergic neurons in the ventral periaqueductal gray matter that expressed Fos protein during natural and environmentally induced W, but none of the neurons expressed Fos during sleep. Fos immunoreactivity was not seen in the SN dopaminergic cells in either condition. Lesion of the periaqueductal gray matter neurons by the injection of 6-hydroxydopamine, which killed 55–65% of wake-active dopaminergic cells but did not injure nearby serotonergic cells, increased total daily sleep by approximately 20%. In addition, these ventral periaqueductal neurons, like the dopaminergic ones in the VTA, are controlled by ventrolateral preoptic nucleus sleep-promoting neurons that contain the inhibitory neurotransmitters GABA and galanin and receive excitatory connections from hypothalamic Hcrt/Orx, DR serotonergic, LdT cholinergic, and LC noradrenergic wake-active neurons (Lu et al. 2006).

It is known that VTA and SN dopamine neurons do not change their mean firing rate over the course of SWC (Miller et al. 1983), although considerable evidence has shown that the dopamine cells may present a change in their temporal firing pattern during the SWC. In fact, a recent study in rats has demonstrated that during behaviorally active W as well as during REM sleep, there is an increase of burst firing activity by ventral midbrain dopaminergic neurons (Dahan et al. 2007). Enhanced dopamine release in a number of basal ganglia and cortical structures has been described as associated with bursting activity by dopamine cells (for review see Monti and Monti 2007). These observations, together with the SWC disturbances displayed by individuals with altered dopaminergic transmission (Abbott 2005; Adler 2005), support the hypothesis of Dzirasa et al. (2006) that dopamine plays an important role in regulating the SWC (Monti and Monti 2007).

However, the difference in prosencephalic dopaminergic innervations between animals as different as rats and primates should not be overlooked: in both cases VTA and the SN innervate telencephalic structures, but whereas in the rat, the thalamus is very scarcely innervated by dopamine, and in the monkey and man, it is densely and extensively innervated by dopaminergic fibers (García-Cabezas et al. 2009). This dopaminergic innervation of the thalamus in primates originates in neurons situated in the hypothalamus, periaqueductal gray matter, ventral midbrain, and the lateral parabrachial nucleus (Snchez-Gonzlez et al. 2005). We emphasize, quoting the words of García-Cabezas et al. (2009), that it is important to be aware of brain species differences when using animal models to study human and other mammal brain "functions".

On the other hand, we cannot ignore the midbrain presence of significant GABAergic formations in the reticular SN or the vigorous GABAergic groups of the periaqueductal gray, a region that widely innervates the W-promoting structures of the dorsal mesopontine tegmentum and forebrain (Llamas and Reinoso-Suárez 1969; Luppi et al. 1998). Also, there are GABAergic midbrain reticular formation afferents from the reticular thalamic nucleus and anterior hypothalamus-basal forebrain sleep-promoting structures. This is interesting, since with their widespread axonal distribution and their wide dynamic range, the GABAergic neurons of the ventral tegmental area have been assigned a role in the regulation of cortical activity as a component of the subthalamic branch of the ascending reticular activating system (Lee et al. 2001).

In summary, the midbrain is an important level in the waking neuronal network because the waking pathways originating in the dorsal mesopontine tegmentum ascend through it on their way to the forebrain, thereby influencing the midbrain structures that, in their turn, collaborate and add their own input to the ascending waking impulses.

3.1.3

The Posterior Lateral Hypothalamus

Eisensehr et al. (2003), studying a patient with a bilateral posterior hypothalamic lesion, confirmed previous observations in humans by von Economo (1926, 1930) and results from animal experiments (see Chap. 2; Nauta 1946; Lindsley et al. 1950; Moruzzi 1972; Reinoso-Suárez and De Andrés 1976); they concluded that the lateral posterior hypothalamus plays a critical role in the maintenance of W in humans. Lin et al. (1989) reached similar conclusions using hypothalamic microinjection of muscimol (a GABA-A agonist) in freely moving cats, finding that the posterior hypothalamus plays a critical role in the mechanisms of W and that sleep might result from a functional blockade of the hypothalamic waking center.

The axons of neurons located in either the midbrain reticular core or in chemically identified cell groups ascend together with axons originating in the pontine tegmentum to enter the diencephalon (Reinoso-Suárez 1977, 2005). Some of these axons, the ones with a midbrain or pontine origin, end in the thalamus, while the rest contribute to the medial forebrain bundle in the lateral hypothalamus that then runs onto the basal and cortical telencephalic formations preserving its precise topographic organization (Pasquier and Reinoso-Suárez 1978). Many of these fibers target posterior lateral hypothalamic and subthalamic cellular groups that are important in the organization of the SWC, specially the zona incerta, the lateral posterior hypothalamic histaminergic tuberomamillary nucleus, and the Hcrt/Orx and the melanin-concentrating hormone neurons (Reinoso-Suárez 1985, 2005; Saper et al. 2005). All of these cell groups join the hypothalamic glutamatergic, dopaminergic, and GABAergic neurons, the latter from both the hypothalamus and the zona incerta, and project onwards to the thalamus (Figs. 2.34 and 2.35). Most of these diencephalic structures project caudally to the brainstem nuclei involved in the organization of the SWC; they also project rostrally, through the medial forebrain bundle, to the basal forebrain, as well as other subcortical telencephalic structures and the cerebral cortex (Fig. 2.42).

The lateral posterior hypothalamic structures play a critical role in the induction and maintenance of proper W. Two neuronal groups, the histaminergic tuberomamillary and the Hcrt/Orx cellular groups, are recognized as playing an important role in organizing the W state.

Histamine. Histaminergic neurons are located in the tuberomammillary nucleus and adjacent areas of the posterior lateral hypothalamus (Fig. 3.3) (Watanabe et al. 1984; Lin et al. 1986; Sallmen et al. 1999). These neurons are important in various physiological functions such as learning, memory, attention, and SWC control (Schwartz et al. 1991; Wada et al. 1991; Haas and Panula 2003; Takahashi et al. 2006).

Studies in cats and rats demonstrated that putatively histaminergic latero-posterior hypothalamic neurons fired tonically at low rates during W, decreased firing in NREM sleep and nearly ceased firing during REM sleep (Vanni-Mercier et al. 1984; Steininger et al. 1999). Takahashi et al. (2006), using extracellular

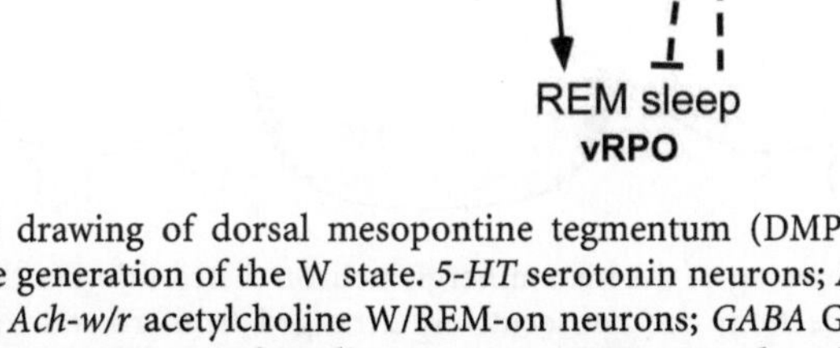

Fig. 3.3 Schematic drawing of dorsal mesopontine tegmentum (DMPT) neuronal types responsible for the generation of the W state. *5-HT* serotonin neurons; *Ach-r* acetylcholine REM-on neurons; *Ach-w/r* acetylcholine W/REM-on neurons; *GABA* GABA neurons; *Glu* glutamatergic neurons; *NA* noradrenalin neurons; *vRPO* ventral oral pontine reticular nucleus. *Dashes* and *arrows*, respectively, represent inhibitory and excitatory connections. For more detail see text

single-unit recordings in combination with neurobiotin juxtacellular labeling and histamine immunohistochemistry in unanesthetized mice, have shown that histamine neurons in the tuberomammillary nuclei are only active during W. The activity of histaminergic neurons varies in relation with the different levels of vigilance, being lowest during quiet W, moderate during active W, and highest during attentive W. These neurons cease firing during the drowsy state and remain silent during NREM and REM sleep. These data support, in their authors' view, the important role of the histaminergic tuberomammillary neurons in the maintenance of the level of vigilance that is necessary for cognitive processes. Histamine release increases on wakening in the posterior hypothalamus of monkeys and this increase is maintained during each waking episode (Onoé et al. 1992).

The histaminergic neurons send axons to diverse brain areas that are important in SWC mechanisms; these areas include the dorsal mesopontine tegmentum, anterior hypothalamus-basal forebrain, thalamus, and cerebral cortex (Figs. 3.4 and 3.5) (Inagaki et al. 1988; Panula et al. 1989; Lin et al. 1993, 1996). It has been suggested that histaminergic neurons might enhance W both by exerting an inhibitory control over sleep-generating structures such as those of the anterior hypothalamus-basal forebrain, as well as by exerting an excitatory action on the structures related to or participating in the induction and maintenance of W. These actions may be mediated by H_1 receptors, which like H_2 receptors mediate excitatory postsynaptic actions in most of these brain structures (Haas and Panula 2003).

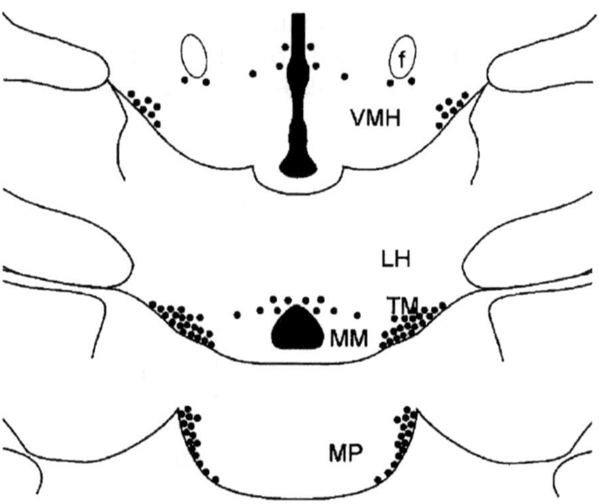

Fig. 3.4 Schematic drawing of the histaminergic neuron distribution in the ground squirrel brain. The frontal sections are consecutively arranged from rostral to caudal. One dot represents approximately two cell bodies. *f* fornix; *LH* lateral hypothalamic area; *MM* medial mammillary nucleus; *MP* posterior mammillary nucleus; *TM* tuberomammillary nucleus; *VMH* ventromedial hypothalamic nucleus. From Sallmen et al. (1999)

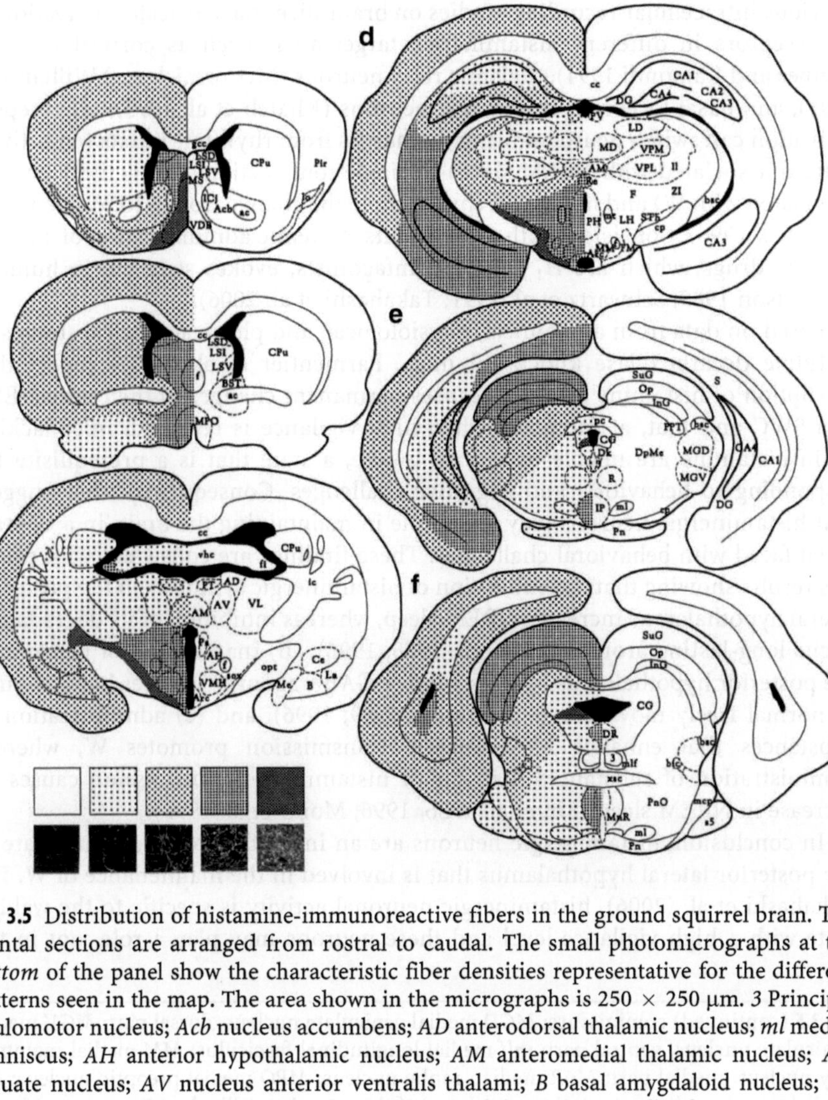

Fig. 3.5 Distribution of histamine-immunoreactive fibers in the ground squirrel brain. The frontal sections are arranged from rostral to caudal. The small photomicrographs at the *bottom* of the panel show the characteristic fiber densities representative for the different patterns seen in the map. The area shown in the micrographs is 250 × 250 μm. *3* Principal oculomotor nucleus; *Acb* nucleus accumbens; *AD* anterodorsal thalamic nucleus; *ml* medial lemniscus; *AH* anterior hypothalamic nucleus; *AM* anteromedial thalamic nucleus; *Arc* arcuate nucleus; *AV* nucleus anterior ventralis thalami; *B* basal amygdaloid nucleus; *bic* brachium of the inferior colliculus; *bsc* brachium of the superior colliculus; *mt* mammillothalamic tract; *BST* bed nucleus of the stria terminalis; *CA1-4* fields CA1–4 of ammons horn; *opt* optic tract; *cc* corpus callosum; *Ce* central amygdaloid nucleus; *pc* posterior commissure; *CG* central gray; *cp* cerebral peduncle; *CPu* caudate putamen (striatum); *DG* dentate gyrus; *Dk* nucleus of Darkschewitsch; *DpMe* deep mesencephalic nucleus; *DR* dorsal raphe nucleus; *F* fields of Forel; *f* fornix; *fi* fimbria; *gcc* genu corpus callosi; *ICj* islands of Calleja; *InG* intermediate gray layer of the superior colliculus; *IP* interpenduncular nucleus; *La* lateral amygdaloid nucleus; *LD* laterodorsal thalamic nucleus; *ll* lateral lemniscus; *lo* lateral olfactory tract; *LSD* lateral septal nucleus, dorsal part; *LSI* lateral septal nucleus, intermediate part; *LSV* lateral septal nucleus, ventral part; *mcp* middle cerebral peduncle; *MD* mediodorsal thalamic nucleus; *Me* medial amygdaloid nucleus; *ac* anterior

Previous intracellular recording studies on brain slices have revealed activation of H_1 receptors in different histaminergic target areas such as cortical neurons (Reiner and Kamondi 1994), thalamic relay neurons (McCormick and Williamson 1991), and basal forebrain cholinergic neurons (Khateb et al. 1995); H_1 receptor activation can switch neuronal firing discharges from rhythmic bursts (an activity pattern associated with NREM sleep) to single tonic activity (an activity pattern associated with W) and might thereby promote the change from NREM sleep to W (Lin et al. 1996). Consistent with these results, systemic administration of antihistaminic drugs, which are H_1 receptor antagonists, evokes sedation in humans (Nicholson 1983; Schwartz et al. 1991; Takahashi et al. 2006).

Based on data from anatomical, physiological, and pharmacological studies of histidine decarboxylase knock-out mice, Parmentier et al. (2002) report that disruption of histamine synthesis causes permanent changes in the cortical-EEG and SWC and that, at moments when high vigilance is required, mice lacking brain histamine are unable to remain awake, a state that is a prerequisite for responding to behavioral and cognitive challenges. Consequently, they suggest that histaminergic neurons play a key role in maintaining the brain in a W state when faced with behavioral challenges. These findings are consistent with previous results showing that: (a) inhibition of histaminergic synthesis in the posterior lateral hypothalamus increases NREM sleep, whereas inhibition of its degradation elicits long-lasting arousal (Lin et al. 1988, 1996); (b) inactivation of neurons in the posterior hypothalamus by muscimol (a GABA agonist) induces hypersomnia in normal freely moving cats (Lin et al. 1989, 1996); and (c) administration of substances that enhance histaminergic transmission promotes W, whereas administration of substances that impair histaminergic transmission causes an increase in NREM sleep (Lin et al. 1986, 1996; Monti et al. 1991).

In conclusion, histaminergic neurons are an important neuronal substrate in the posterior lateral hypothalamus that is involved in the maintenance of W. For Takahashi et al. (2006), histaminergic neuronal activity is specific to the waking state with a high vigilance level and these neurons may play a role, not in the

Fig. 3.5 (continued) commissure; *MGD* medial geniculate nucleus, dorsal part; *MGV* medial geniculate nucleus, ventral part; *mlf* medial longitudinal fasciculus; *MM* medial mammillary nucleus, medial part; *MnR* median raphe nucleus; *MPO* medial preoptic nucleus; *MS* medial septal nucleus; *Op* optic nerve layer of the superior colliculus; *Pa* paraventricular hypothalamic nucleus; *PH* posterior hypothalamic nucleus; *Pir* piriform cortex; *Pn* pontine nuclei; *PnO* pontine reticular nucleus, oral part; *PT* paratenial thalamic nucleus; *PV* paraventricular thalamic nucleus; *R* red nucleus; *Re* reuniens thalamic nucleus; *S* subiculum; *s5* sensory root of the trigeminal nerve; *sox* supraoptic decussation; *ic* internal capsule; *STh* subthalamic nucleus; *SuG* superficial gray layer of the superior colliculus; *TM* tuberomammillary nucleus; *VDB* nucleus of the ventral limb of the diagonal band of Broca; *vhc* ventral hippocampal commissure; *VL* ventrolateral thalamic nucleus; *VMH* ventromedial hypothalamic nucleus; *VPL* ventral posterolateral thalamic nucleus; *VPM* ventral posteromedial thalamic nucleus; *xsc* decussation of the superior cerebellar peduncle; *ZI* zona incerta. From Sallmen et al. (1999)

initiation of W per se, but in the maintenance of the high level of vigilance necessary for cognitive processes; that is to say, histaminergic neurons play a key role in maintaining the brain in a W state when faced with behavioral challenges. Conversely, cessation of their activity may play an important role in both the initiation and maintenance of sleep. It seems to be that the rich GABAergic innervations of histaminergic neurons by the ventral preoptic nucleus are closely linked to the transition between W and NREM sleep (Saper et al. 2005).

Hypocretins/Orexins. Another important waking hypothalamic group is comprised by the hypocretin/orexin (Hcrt/Orx) containing neurons that strongly participate in the mechanisms of W–S regulation and in the pathophysiology of narcolepsy/cataplexy (Chemelli et al. 1999; Lin et al. 1999; Peyron et al. 2000; Takahashi et al. 2006).

The hypocretins, also called orexins, are two excitatory neuropeptides (Hcrt-1 and -2 or Orx-A and -B) (de Lecea et al. 1998; Sakurai et al. 1998) that are synthesized by neurons located in the posterior, lateral, dorsal and perifornical hypothalamic areas, and in the zona incerta (Fig. 3.6); except for the cerebellum they project to many widespread regions of the central nervous system (Figs. 3.5 and 3.7) (Peyron et al. 1998; Marcus et al. 2001; Lee et al. 2005a; Wang et al. 2005; Torterolo et al. 2006; Sakurai 2007). This projection is densest in the hypothalamus, principally in the arquate and tuberomammillary nuclei and in the locus

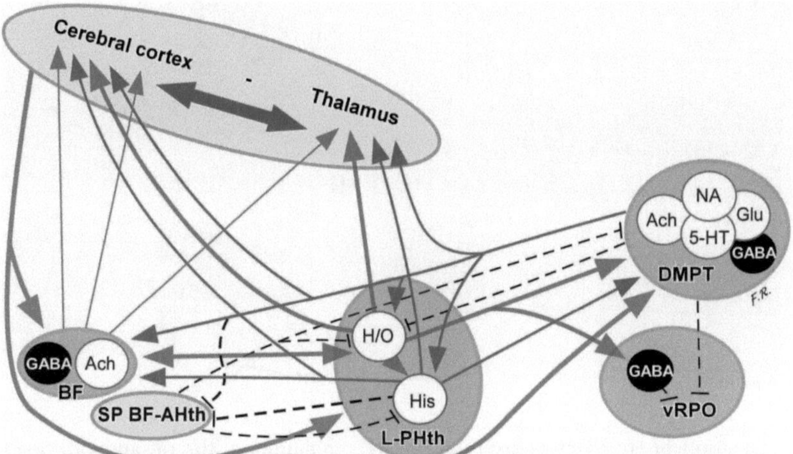

Fig. 3.6 Schematic drawing of the neuronal network responsible for generating and maintaining W. *5-HT* serotonin neurons; *Ach* acetylcholine neurons; *BF* basal forebrain; *DMPT* dorsal mesopontine tegmentum; *GABA* GABA neurons; *Glu* glutamatergic neurons; *H/O* hypocretin/orexin neurons; *His* histamine neurons; *L-PHth* lateral-posterior hypothalamus; *NA* noradrenalin neurons; *SP BF-AHth* sleep-promoting basal forebrain-anterior hypothalamic region; *vRPO* ventral oral pontine reticular nucleus. *Dashes* and *arrows*, respectively, represent inhibitory and excitatory connections. For more detail see text

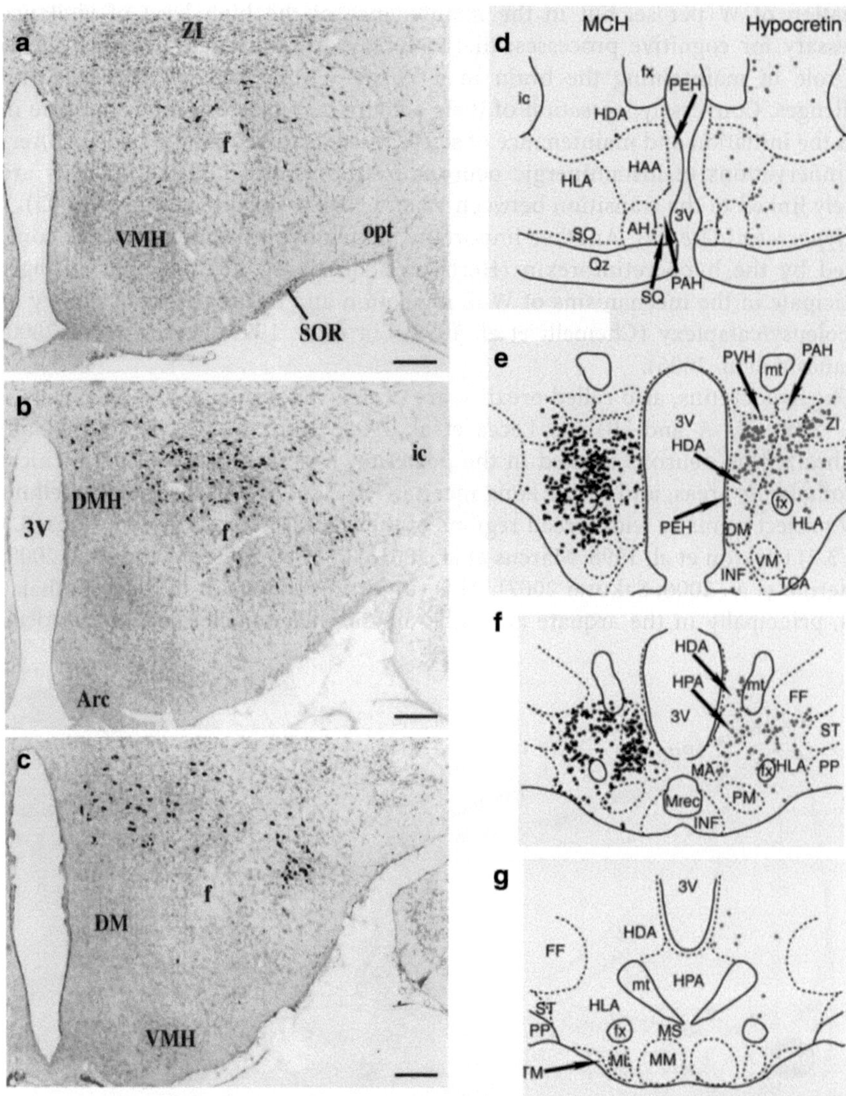

Fig. 3.7 Location of Hctr/Orx neurons in the hypothalamus of the rat and cat. (**a–c**) The distribution of the rat Hcrt/Orx-labeled neurons on frontal sections at three rostrocaudal levels of the tuberal region of the hypothalamus has been determined using #2123 antiserum. Sections were counterstained with neutral *red* (*pink* staining of all cells). *3V* 3rd ventricle; *Arc* arquate nucleus; *DMH, DM* dorsomedial hypothalamic nucleus; *f* fornix; *ic* internal capsule; *opt* optic tract; *SOR* retrochiasmatic part of the supraoptic nucleus; *VMH* ventromedial hypothalamic nucleus; *ZI* zona incerta. *Scale bars*, 275 mm. From Peyron et al. (1998). (**d–f**) Camera lucida drawings of melanin-concentrating hormone (MCH, on the *left*, *black circles*) and hypocretinergic neuronal bodies (on the *right*, *red circles*) at different levels of the cat hypothalamus are shown. The neurons are from the same

coeruleus. Also, it is especially dense in the paraventricular thalamic nucleus, the ventral tegmental area, and the DR. Less dense projections are seen in the septal nuclei, the bed nucleus of the stria terminalis, the paraventricular and reuniens nuclei of the thalamus, the zona incerta, the subthalamic nucleus, the central gray matter, the substantia nigra, the parabrachial area, the medullary reticular formation, and the nucleus of the solitary tract (Fig. 3.7). Consistent projections were found in cortical regions, central and anterior amygdaloid nuclei, and the olfactory bulb. In vitro and in vivo studies have shown that Hcrt/Orx activate cells in most of these nuclei.

Lee et al. (2005a) recorded neurons in head-fixed rats over the SWC and labeled them with neurobiotin to identify them. They demonstrated that identified Hcrt/Orx neurons discharge during active W, decrease their discharge rate during quiet W and virtually cease firing during REM and NREM sleep. The neurons increase their firing before the end of REM sleep, heralding the return of waking and muscle tone by several seconds. Those authors conclude that the Hcrt/Orx neurons would stimulate arousal while antagonizing sleep and muscle atonia. Summing up, these and similar results from Mileykovskiy et al. (2005) are solid evidence that Hcrt/Orx neurons are active during W and silent during sleep. The Hcrt/Orx neuron discharge is positively correlated with EMG amplitude, and their increased firing prior to the return of muscle tone with awakening from REM sleep appears to stimulate awakening by recruiting other arousal systems (Jones 2008).

Hcrt/Orx neurons receive and are modulated by excitatory and inhibitory inputs. The excitatory susbstances are glutamate, ghrelin, cholecystokinin, neurotensin, vasopressin, oxytocin, glucagon-like peptide, corticotropin-releasing factor, mACh (affecting 27% of all Hcrt/Orx neurons), and ATP; the inhibitory susbstances are glucose, GABA, serotonin, noradrenaline, dopamine, neuropeptide Y, leptin, mACh (affecting 6% of Hcrt/Orx neurons), and adenosine (see Tsujino et al. 2005; Sakurai 2007). These impulses implicate the innervation of Hcrt/Orx neurons in a large number of brain structures that, in some cases, as Yamanaka et al. (2006) specify studying Hcrt/Orx neuron regulation by catecholamines, may be innervated directly or indirectly through other structures, most normally in a complex manner. Effectively, a feedback loop between Hcrt/Orx neurons and the LC and DR monoaminergic neurons might maintain the activity of the latter; decreases in monoaminergic neuron activity will decrease the inhibitory influence on Hcrt/Orx neurons, this disinhibition of Hcrt/Orx neurons would then increase the excitation of monoaminergic cells, thereby increasing W (Fig. 3.5). It has not been easy to pinpoint the afferent connections of the

Fig. 3.7 (continued) hemi-hypothalamus (reflected in the figure). Both types of neurons were intermingled and were densest at the tuberal level. Camera lucida drawings were obtained from adjacent sections, one was immunostained for MCH and the other was immunostained for Hcrt-2; these sections were counterstained with Pyronin-Y. Modified from Torterolo et al. (2006)

Hcrt/Orx neurons because they are dispersed over a wide hypothalamic region. In addition, monoaminergic and peptidergic fibers may use "volume transmission," which includes diffusion of signals through the extracellular and cerebrospinal fluid. Actually, tracing studies showed that projections of monoaminergic neurons to Hcrt/Orx neurons are sparse and that Hcrt/Orx neurons are apposed to NPY-containing peptidergic fibers. Also, extrasynaptic receptors can sense ambient ligands, which can act as neuromodulators (Sakurai 2007).

Recently, many of these afferents have been traced using a genetically encoded retrograde tracer or with a combination of anterograde and retrograde tracers (Sakurai et al. 2005; Yoshida et al. 2006). In summary, the Hcrt/Orx neurons receive abundant inputs from the lateral septum, preoptic area, including forebrain cholinergic neurons and GABAergic neurons of the peoptic area, and the posterior hypothalamus; consistent input from the the allocortex, claustrum, amygdala, basal bed nucleus of the stria terminalis and from many hypothalamic regions including the dorsomedial nucleus and lateral hypothalamus also reaches these neurons (Fig. 3.8). The number of afferents from the brainstem is, in general, more moderate, the strongest coming from the periaqueductal gray matter, DR nucleus, and lateral parabrachial nucleus. Monoamine-containing groups that are excited by Hcrt/Orx neurons reciprocally inhibit Hcrt/Orx

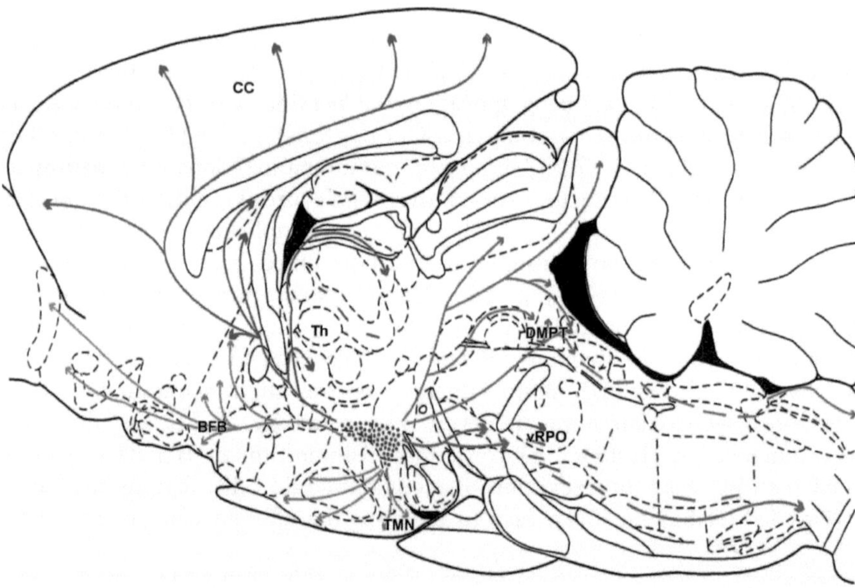

Fig. 3.8 Schematic summary drawing of the Hcrt/Orx pathways that widely innervate the rat brain. *Purple dots*: Hcrt/Orx-labeled neurons; *red*: dorsal ascending pathway; *light blue*: ventral ascending pathway; *green*: dorsal descending pathway; *dark blue*: ventral descending pathway. From Peyron et al. (1998)

neurons, while cholinergic neurons in the basal forebrain have reciprocal excitatory connections with Hcrt/Orx neurons, which might be important for consolidating W (Fig. 3.5). In addition, hypothalamic regions preferentially innervate Hcrt/Orx neurons situated in the medial and perifornical regions, but most projections from the brainstem target Hcrt/Orx neurons located in more lateral areas. The preoptic area and the ventrolateral preoptic nucleus have been considered a source of GABAergic inhibitory projections to Hcrt/Orx neurons. These pathways might be important for inhibiting Hcrt/Orx neurons during sleep (Fig. 3.5). The latter two sleep-promoting centers are also innervated by inhibitory neurotransmitters, mainly GABA, from other W promoting structures such as the tuberomamillary nucleus and are inhibited directly by the serotonergic neurons of the DR and the noradrenergic neurons of the LC (Fig. 3.5) (Sherin et al. 1998). Finally, Hcrt/Orx neurons may be innervated indirectly, receiving circadian influences from the suprachiasmatic nucleus given the projections of this nucleus to the basal bed nucleus of the stria terminalis and the dorsomedial hypothalamic nucleus and from these nuclei to the Hcrt/Orx neurons.

In relation with the caudal projections of the Hcrt/Orx neurons to mesopontine tegmentum structures related with the organization of the SWC, Burlet et al. (2002) suggested that Hcrt/Orx neurons activate both monoamine and cholinergic components of the ascending reticular activating system to promote EEG desynchronization and W (Figs. 3.5 and 3.7). Indeed, focal application of Hcrt/Orx in the noradrenergic LC (Bourgin et al. 2000) and in cholinergic LdT in the cat (Xi et al. 2001) produces arousal. What is more, glutamatergic afferents to LdT neurons are strongly stimulated by Hcrt/Orx (Burlet et al. 2002); therefore, Hcrt/Orx neurons provide excitatory input to the glutamatergic neurons of the reticular formation, as well. According to Burlet et al. (2002), hypothalamic Hcrt/Orx neurons may coordinate activation of the entire ascending reticular activating system. In addition, Hcrt/Orx directly excites histaminergic tuberomamillar neurons as demonstrated in vitro in rat brain slice preparations (Yamanaka et al. 2002). Studies in vivo in rat and mice have shown that perfusion of Hcrt 1 into the tuberomamillar nucleus of rats through a microdialysis probe promptly increases W and histamine release, while the intracerebroventricular injection of Hcrt 1 selectively activates the histaminergic system and causes a significant increase in W in wild-type mice but not in histamine H1 receptor gene knockout mice (Huang et al. 2001); the conclusion was that the arousal effect of Hctr 1 depended on the activation of histaminergic neurotransmission mediated by histamine H1 receptor.

Recently, our group has demonstrated that low-volume microinjections of Hctr1 in the dorsal mesopontine tegmentum produce an increase of W that was associated with decrease of both NREM and REM sleep in behaving cats (Moreno-Balandrn et al. 2008). Since Hctr-1 produces excitatory effects in all the cell types in this region (Brown et al. 2006), these W promoting effects would probably have been produced by Hctr-1 excitation of the cholinergic, monoaminergic, glutamatergic, and GABAergic neurons present in the dorsal mesopontine tegmentum.

The sleep decrease might be more than a simple consequence of the W increase, since the excitated dorsal mesopontine tegmentum waking neurons could inhibit the forebrain NREM sleep-promoting structures (Détri et al. 1997; Gallopin et al. 2000; Chou et al. 2002) as well as the pontine REM sleep generator (Nuñez et al. 1998; Rodrigo-Angulo et al. 2000) (Fig. 3.5) (Moreno-Balandrn et al. 2008).

Nevertheless, Hcrt/Orx hypothalamic neurons project to the REM sleep generating ventral oral pontine nucleus (vRPO) and iontophoretic application of Hcrtl inhibits vRPO neurons through GABA-A receptor activation (Fig. 3.5) (Nuñez et al. 2006), so it could be possible that Hcrt/Orx neurons directly inhibit pontine REM sleep-promoting neurons. Indeed, Hcrtl microinjections in vRPO produce a direct and selective inhibition of REM sleep (Moreno-Balandrn et al. 2008).

Thus, all these experimental findings show that the Hcrt/Orx neurons have an important role in maintaining W by acting on brain areas implicated in arousal, but they also block the generation of REM sleep by acting directly on its generator (Nuñez et al. 2006; Moreno-Balandrn et al. 2008). Therefore, in narcolepsy, the loss of Hcrt/Orx signaling would de-facilitate the entire waking ascending reticular activating system and remove the inhibitory action of Hcrt/Orx on the vRPO REM sleep generator, consequently facilitating the narcoleptic patients' characteristic of abrupt REM sleep onset during a W period.

The rostral projection of the Hcrt/Orx neurons to basal forebrain is also important for inducing and maintaining W: administration of Hcrt/Orx in basal forebrain induces W in rats. Hcrt/Orx depolarizes basal forebrain cholinergic neurons (Eggermann et al. 2001; España et al. 2001; Thakkar et al. 2001) and destruction of over 90% of the Hcrt/Orx neurons by hypothalamic injection of the neurotoxin Hcrt-2-saporin decreases W and increases sleep (Fig. 3.5) (Murillo-Rodriguez et al. 2008).

Adamantidis et al. (2007) found that direct, selective, and optogenetic photostimulation of Hcrt/Orx neurons using 5–30 Hz light pulse trains significantly reduced latencies to W, increasing the probability of transition to W from either NREM sleep or REM sleep. These findings led those authors to state that the Hcrt/Orx neurons have a role in initiating the transition from sleep to W.

Other findings point to an Hcrt/Orx role for the maintenance of waking and behavioral arousal. The role of Hcrt/Orx in maintaining the level of W has been demonstrated in experiments measuring the daily time pattern of hypocretin-1 content in the cysterna magna of the squirrel monkey (Zeitzer et al. 2003). This New World primate has a W pattern that is similar of the one in humans. In the squirrel monkey, Hcrt-1 remains low during the initial 1–3 h of W, despite elevated locomotor activity present from the onset of W in these animals. This argues against Hcrt/Orx being necessary for the expression of W during these early hours of the day and consequently Hcrt/Orx may not be necessary for initiating the transition from sleep to W. In addition, Hcrt-1 levels in the squirrel monkey peaked in the last third of the day and, if the W period was experimentally lengthened by 4 h, the Hcrt-1 concentrations remained elevated (Zeitzer et al. 2003).

These results indicate both that Hcrt-1 is involved in W regulation and that there is a circadian-independent component to hypocretin-1 regulation. The Zeitzer group concludes that in the squirrel monkey, Hcrt-1 works in opposition to the accumulating sleep drive during the day to maintain a constant level of W. Hcrt/Orx neurons also participate in other functions such as feeding behavior, energy homeostasis, stress response, the reward system, and the autonomic nervous system (Sakurai 2007; Nuñez et al. 2009).

Other hypothalamic neurotransmitters. Another widely distributed hypothalamic neuronal group that has been related with SWC control are the melanin-concentrating hormone (MCH) neurons that Verret et al. (2003) described as being strongly active during REM sleep. Torterolo et al. (2006) showed that MCH neuronal somata in the cat are intermingled with Hcrt/Orx neurons in the dorsal and lateral hypothalamus, mainly in the tuberal and tuberomammillary regions (Fig. 3.6), and observed that axosomatic and axodendritic contacts were common between these neurons. It is also known that Hcrt/Orx has a profound excitatory effect on MCH neurons (Bayer et al. 2002a; van den Pol et al. 2004). Still, c-fos immunoreactivity was not observed in MCH-containing neurons in conjunction with any of the behavioral conditions of an extensive battery of conditions related with the SWC explored in the cat by Torterolo et al. (2006). However, a neuronal group with an unidentified phenotype expressed c-fos during active W with motor activity and, this group was found to be interspersed with Hcrt/Orx and MCH neurons, suggesting that these neurons are related to some aspect of motor function.

Besides neurons characterized biochemically in the lateral posterior hypothalamic waking region, there is another large number of neurons that may contribute to W maintenance and that project to brainstem, basal forebrain, cerebral cortex, and thalamus whose neurotransmitters are not yet identified. Possibly, most of these neurons use glutamate or other excitatory and/or inhibitory (principally GABA) neurotransmitters or neuromodulators. As an example, De la Roza et al. (2004) and Rodrigo-Angulo et al. (2008) have demonstrated GABAergic inhibitory projections to the vRPO (REM sleep induction site), from the dorsal and lateral hypothalamus in areas in which a large proportion of the neurons are most active during W (Lin 2000; Saper et al. 2005; Sutcliffe and de Lecea 2002; Koyama et al. 2003); therefore, this GABAergic projection may contribute to W maintenance by inhibiting the REM sleep generating neurons. On the other hand, De la Roza et al. (2004) report that a large number of the non-GABAergic hypothalamic terminals on the vRPO neurons form symmetric synapses. These putatively inhibitory terminals may also originate in the hypothalamic neurons that are most active during W (Koyama et al. 2003; Torterolo et al. 2006) and use some of the many non-GABA hypothalamic neurotransmitters to inhibit vRPO neurons during W (Rodrigo-Angulo et al. 2008).

Summary: The posterior lateral hypothalamus completes the structures that compose the "Ascending Reticular Activating System," the principal system that is responsible for the W phase in the SWC (Fig. 3.9). We now know the main structures in this system, their main neurotransmitters, their interconnections and

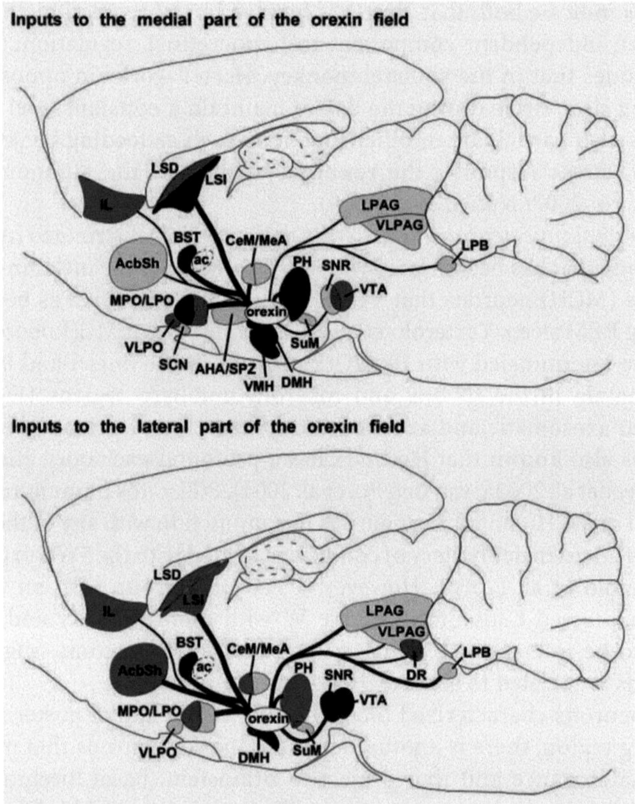

Fig. 3.9 Summary of afferent projections to the orexin neurons. The *upper panel* shows inputs to orexin neurons in the medial part of the orexin field and the *lower panel* shows inputs to orexin neurons in the lateral part of the field. The heaviest inputs are from the preoptic area, dorsomedial nucleus, posterior hypothalamus, lateral septum, and bed nucleus of the stria terminalis. This pattern suggests that the orexin neurons are influenced by signals regulating emotions, autonomic tone, appetite, circadian rhythms, and sleep/ wake behavior. Regions labeled in *dark*, *medium*, and *light gray* innervate >45%, 25–44%, or 5–24% of the orexin neurons, respectively. Inputs that innervate <5% of the orexin neurons are not included. *Line thickness* indicates the relative number of retrogradely labeled neurons. *ac* anterior commissure; *AcbSh* nucleus accumbens; *shell AHA* anterior hypothalamic nucleus; *Arc* arcuate nucleus; *BST* bed nucleus of the stria terminalis; *DMH* dorsomedial nucleus of the hypothalamus; *DR* dorsal raphe nucleus; *IL* infralimbic cortex; *LC* locus coeruleus; *LPB* lateral parabrachial nucleus; *LPO* lateral preoptic area; *LSD* lateral septum, dorsal part; *LSI* lateral septum, intermediate part; *MPO* median preoptic area; *PAG* periaqueductal gray; *PH* posterior hypothalamus; *SCN* suprachiasmatic nucleus; *SNR* substantia nigra, reticular part; *SPZ* subparaventricular zone; *SuM* supramammillary nucleus; *TMN* tuberomammillary nucleus; *VLPO* ventrolateral preoptic area; *VMH* ventromedial nucleus of the hypothalamus; *VTA* ventral tegmental area; *Z* zona incerta. From Yoshida et al. (2006)

connections and the influences that they exert on or receive from other brain formations, essentially those related with the organization of the SWC. In this section, we have emphasized the role of two important purely hypothalamic cellular groups in the correct emergence of W: the histamine and Hcrt/Orx-containing neurons. These neurons interact and add their influence to the serotonergic, noradrenergic, cholinergic, dopaminergic, glutamatergic, and GABAergic waking neurons of the midbrain and dorsal mesopontine tegmentum to attain an efficient W state (Figs. 3.5 and 3.9). This complex system, primarily using glutamate as a neurotransmitter, in addition to its action on the brainstem, cerebellum, and spinal cord, promoting motor activity with postural muscle tone, would awaken the thalamus and cerebral cortex directly or indirectly through the basal forebrain waking neurons, promoting desynchronized bioelectrical activity. At the same time, the "ascending reticular activating system" would inhibit the brain-sleep inducing structures during the W periods of the SWC, and it would be reciprocally inhibited by these sleep-inducing structures during the sleep phases.

3.2 Other Brain Structures with Their Neurotransmitters That Participate in Wakefulness

The final targets of the "ascending reticular activating system" are, as classically held, the thalamus and the cerebral cortex. The system projects directly on both the thalamus and cortex with its two branches: one thalamic, ending in the thalamus, the other subthalamic, ending in the cerebral cortex. It also projects on and is influenced by the basal forebrain waking neurons that modulate the activity of these brain structures through thalamic and cortical projections (Figs. 3.5 and 3.9). We will divide the study of the final waking action of the reticular activating system into two subsections: 3b1, anterior hypothalamic and basal forebrain regions and 3b2, thalamus and cerebral cortex.

3.2.1 Anterior Hypothalamic and Basal Forebrain Regions

Chapter 2 emphasized the importance of the basal forebrain-anterior hypothalamic region in NREM sleep induction. The previous sections of Chap. 3 have described inhibitory influences during W from the wakefulness promoting structures of the brainstem and lateral posterior hypothalamus on the basal forebrain sleep-promoting regions such as the preoptic area and the ventrolateral preoptic nucleus; in their turn, these sleep-promoting formations inhibit the waking promoting structures of the brainstem and lateral posterior hypothalamus during sleep (Fig. 3.5). We have also noted that GABA is the main neurotransmitter of these sleep-promoting areas. Basal forebrain GABAergic sleep-active neurons are intermingled with the basal forebrain cholinergic and GABAergic waking neurons that project to the cerebral cortex. The sleep-active cells discharge maximally in association with slow wave cortical activity and minimally in association with fast wave cortical activity (Manns et al. 2000). Noradrenalin, an important component of the reticular activating system, coming from the LC, excites the cholinergic waking basal forebrain neurons and inhibits the slow wave sleep-active GABAergic basal forebrain neurons.

Most of the axons of the waking posterior hypothalamic neurons join the medial forebrain bundle in the lateral hypothalamus where, together with those ascending from the brainstem, they continue on their way to the basal and cortical telencephalic formations. This is the pathway used by most of these ascending activating and descending inhibitory connections mentioned in the previous paragraphs. But, in addition to sleep-promoting cell groups, the basal forebrain has cell groups that play an important role in cortical arousal. The basal forebrain contains cholinergic and noncholinergic neurons, most of these being GABAergic but others are glutamatergic, and all project to the cerebral cortex and thalamus (Velayos and Reinoso-Suárez 1985; Steriade et al. 1987b; Gritti et al. 1997; Jones 2007; Henny and Jones 2008). The basal forebrain waking neurons are situated in the basal magnocellular nucleus, innominate substance, Broca's diagonal band, and medial septum. Most of the cholinergic fibers of the ascending activating system that originate in the dorsal mesopontine tegmentum make connections with basal forebrain W promoting cholinergic neurons that project, with a precise topographic organization, to the cerebral cortex and thalamus (Figs. 2.34, 2.42, 3.5, and 3.9) (Kievit and Kuypers 1975a, b; Reinoso-Suárez et al. 1982; Velayos and Reinoso-Suárez 1985; Steriade et al. 1987a, b; Mesulam and Geula 1988; Heckers et al. 1992; Jones 2007). Cholinergic projections stimulate fast gamma activity (30–60 Hz) in the cerebral cortex during W (Jones 2008). A small number of brainstem cholinergic fibers project directly to the cerebral cortex, and many of them excite the basal forebrain cholinergic neurons. These neurons are also excited by glutamate fibers, as well as by noradrenergic fibers, from the reticular activating system and they discharge tonically, giving rise to an increased acetylcholine release in the cortex and thalamus that also produces a shift from slow wave activity to a gamma activity characteristic of cortical activation and behavioral W. In single-unit recording studies, basal forebrain cholinergic neurons have been found to discharge at their highest rates in association with gamma activity during active W and REM sleep and cease discharge during NREM sleep (Alonso 1998; Lee et al. 2005b). However, serotonin inhibits the basal forebrain cholinergic neurons and, in consequence, modulates cortical activation. Reticular activating system terminals in the basal forebrain may release glutamate, which drives specific cholinergic neurons that could stimulate rhythmic theta frequency activity, as also occurs in the hippocampus in W and REM sleep. In W, the basal cholinergic neurons can stimulate gamma and theta activity in the cerebral cortex, including the cingulate cortex, which is associated with vivid attention or motor processing.

Basal forebrain cholinergic neurons are excited by ascending activating influences and produce a similar cortical gamma activity during cortical activation in both W and REM sleep; however, there are important anatomical and functional differences in both cases: (1) in REM sleep, the excitation of the basal cholinergic neurons is produced not only by the projections of the W/REM-on mesopontine cholinergic neurons (which also participate in W), but also by the REM-on mesopontine cholinergic neurons; and (2) in W, basal cholinergic neuron excitation is produced by excitation of W/REM-on mesopontine cholinergic neurons, which exhibit the highest firing rates in W (Kayama et al. 1992) as well as by noradrenergic mesopontine and histaminergic and Hcrt/Orx hypothalamic neurons, which are silent in REM sleep. Also, in W, the basal forebrain neurons are modulated by the active serotonergic neurons (Thakkar et al. 1998). The glutamatergic, GABAergic, and other ascending influences, feasibly different in W and REM sleep [glutamatergic fibers from the RPO nucleus may also participate in the latter phase (Reinoso-Suárez et al. 1990; Reinoso-Suárez et al. 1994)], may mark the behavior of the basal forebrain neurons in each case. As we have described in previous paragraphs, the basal forebrain cholinergic waking neurons have reciprocal excitatory connections with Hcrt/Orx hypothalamic neurons, which might be important for consolidating W (Fig. 3.5). However, the activity of the GABAergic basal forebrain sleep-active neurons suppresses activity by the Hcrt/Orx wake-active neurons and promotes sleep (Murillo-Rodriguez et al. 2008).

The reticular activating system produces gamma activity, characteristic of cortical activation during W, not only through the activating influence of the cholinergic basal forebrain neurons but possibly using other unidentified neurons, which are intermingled with and project toward the cortex in parallel with the cholinergic basal forebrain neurons, such as GABAergic or glutamatergic neurons. These noncholinergic wake-promoting neurons may also be excited by noradrenaline and inhibited by serotonin and would consequently act in concert with cholinergic basal forebrain neurons to produce cortical high-frequency gamma activity.

The basal forebrain cortical projecting fibers join the ascending activating fibers originated in the hypothalamus and brainstem, most of them located in the medial forebrain bundle, enter the subcortical telencephalic structures and, using the internal and external capsule, the cingular fascicle, and the fornix-fimbria, end with a specific pattern for each fiber type, in the cerebral neo- and allo-cortex (Llamas and Reinoso-Suárez 1969; Reinoso-Suárez and Llamas 1975; Reinoso-Suárez 1977; Pasquier and Reinoso-Suárez 1978).

3.2.2
Thalamus and Cerebral Cortex

The thalamus and the cerebral cortex are the final target of the two branches of the reticular ascending activating system: the thalamus of the dorsal or thalamic branch and the cerebral cortex of the ventral or subthalamic branch. The two structures also receive connections from the basal forebrain waking structures (Figs. 3.5 and 3.9). But thalamus and cerebral cortex are widely and complexly interconnected anatomically, phylogenetically, and functionally, as we have described in Chap. 2 (Figs. 2.32, 2.33, 2.40, 2.41, 3.5, and 3.9). Therefore, the behavior of the thalamus and cerebral cortex is closely coordinated in relation with the organization of the SWC, and specifically with the W state.

In Chap. 2 we have described, using retrograde tracer experiments in the cat, thalamic and cortical afferents from the brainstem, hypothalamus, and basal forebrain as W-related structures (Llamas et al. 1975; Pasquier and Reinoso-Suárez 1976, 1977, 1978; Reinoso-Suárez et al. 1982; Rodrigo-Angulo and Reinoso-Suárez 1982; Velayos and Reinoso-Suárez 1982, 1985; Jiménez-Castellanos and Reinoso-Suárez 1985; Clasc et al. 1989); thus, cholinergic, aminergic, and peptidergic fibers arising from these subcortical structures have been identified in the thalamus and cerebral cortex (Figs. 2.20, 2.21, 2.34, 2.35, 2.41, and 2.42).

The distribution patterns of these fibers and terminals are characteristic of each type of fiber but also differ between species. What is more, each neurotransmitter has a specific distribution in each cortical region, different in every region, and different from the distribution of the other neurotransmitters (see Foote and Morrison 1987 and Lewis et al. 2001). As an example: the different cortical areas exhibit specific regional and laminar specializations of the neurotransmitter innervations; thus, in monkeys, the visual region of the inferotemporal cortex was found to be very lightly innervated by noradrenergic fibers but very densely innervated by serotonergic fibers. In contrast, area 7 of the parietal lobule in these animals was more densely innervated by noradrenergic fibers and less densely innervated by serotonergic fibers than any other visual cortical region examined by Morrison and Foote (1986). Sometimes the distribution of two neurotransmitters is complementary, regionally, or tangentially; thus, for example, noradrenergic fibers are abundant in the deep layers of the visual cortex but are absent in layer IV; while, on the contrary, serotonergic fibers are abundant in layer IV, where it appears that they essentially innervate spiny stellate cells. Equally, the different neurotransmitters of the ascending reticular activating system may activate specific cortical neuron receptors situated in different layers of the cerebral cortex and have diverse functional effects in these neurons (McCormick et al. 1993). Differences in neurotransmitter distribution have also been described between the two brain hemispheres. So, in humans, it appears that terminals of noradrenergic fibers are most abundant in the right hemisphere, while dopaminergic and cholinergic fibers are more numerous in the left hemisphere (Davidson et al. 1992). Similar considerations can be made about distribution patterns for the different types of fibers and terminals of the reticular activating system in the thalamus.

Next we will describe the distribution and functional properties of the main neurotransmitters/neuromodulators in the thalamus and cerebral cortex.

Acetylcholine. The thalamic cholinergic fibers originate in the dorsolateral and pedunculopontine tegmental areas (Ch5 and Ch6 according to Mesulam 1995), as can be presumed from the description of the previous sections, and have been demonstrated by double-labeling experiments using retrograde tracers and immunocytochemistry for choline acetyltransferase (Sofroniew et al. 1985; Hallanger et al. 1987; Steriade et al. 1988). These brainstem cholinergic neurons also colocalize nitric oxide synthase, and, consequently, it is supposed that the cholinergic thalamic innervation is the principal source for nitric oxide in the thalamus (Bickford et al. 1993; Vincent and Hope 1992; Jones 2007). Cholinergic fibers of the dorsal lateral geniculate and, to a lower degree, of the pulvinar thalamic nuclei may originate in the midbrain parabigeminal nuclei (Harting et al. 1991; Bickford et al. 2000). The thalamus, essentially the reticular thalamic nucleus, also receives cholinergic projections from the basal forebrain (Velayos and Reinoso-Suárez 1985; Steriade et al. 1987a, b; Heckers et al. 1992; Jones 2007).

The abundant and extensive cholinergic thalamic innervation is densest in the anterior, intralaminar, pulvinar, and dorsal lateral geniculate nuclei of the dorsal thalamus, in the reticular and ventral lateral geniculate nuclei of the ventral thalamus and the epithalamic medial habenular nucleus, with a pattern that is very like, but not identical, to that of acetylcholinesterase staining; this pattern is different in every species studied so far (Graybiel and Berson 1980; Levey et al. 1987; Fitzpatrick et al. 1989; Wilson et al. 1999).

Electrical stimulation of Ch5 causes a hyperpolarization of GABAergic neurons in the reticular thalamic nucleus. Mesulam (1995) considers that the net effect of Ch5 stimulation is to disinhibit thalamic relay nuclei because of the thalamic reticular nucleus neurons have an inhibitory effect on thalamic relay neurons (Steriade et al. 1990). Nevertheless, the cholinergic terminals form synapses preferentially with proximal dendrites not only from reticular thalamic nucleus neurons but also those from relay cells and inhibitory interneurons of dorsal thalamic nuclei, although this pattern may change in the different nuclei of the dorsal thalamus (de Lima et al. 1985; Patel et al. 1999). No conspicuous morphological differences in these synapses have been observed, and it has been interpreted that the differential modulation on the relay neurons and GABAergic neurons would essentially depend on the receptor type used in each case. Acetylcholine excitation of relay neurons is mediated by the activation of nicotinic (the rapid initial depolarization) and muscarinic (the following slow, longer-lasting depolarization) receptors (Curró-Dossi et al. 1991; McCormick 1992). Nonetheless, the cholinergic inhibition of GABAergic dorsal thalamic interneurons and reticular thalamic nucleus neurons is mediated by the activation of M2 muscarinic receptors, which causes the hyperpolarization of these neurons (McCormick and Pape 1988; Lee and McCormick 1995; Pape and McCormick 1995). Together, the depolarization of thalamic relay cells and the hyperpolarization of GABAergic interneurons and GABAergic reticular thalamic neurons allow the cholinergic system to exert a powerful facilitatory influence over the transfer of thalamic information to the cerebral cortex during W; this influence switches the firing mode of the thalamic relay neurons from the burst mode to the single spike mode (Ben-Ari et al. 1976; McCormick 1992; Steriade 2004).

The reticular and mediodorsal thalamic nuclei in rats, cats, and primates, and also the anterior and the ventromedial nuclei in the cat, receive predominantly ipsilateral afferents from the basal forebrain; these fibers are mostly cholinergic, GABAergic, and glutamatergic but can also use other neurotransmitters (Asanuma 1997). Most of the basal forebrain cholinergic fibers target the thalamic reticular nucleus, while the projections to other thalamic nuclei are poor compared with those originating in the dorsal mesopontine tegmentum (Hallanger et al. 1987). About two-thirds of the abundant cholinergic innervation of the thalamic reticular nucleus originates in the basal forebrain. Acetylcholine and glutamate hyperpolarize the thalamic reticular neurons and so inhibit them (McCormick 1992; Cox and Sherman 1999).

The rich cholinergic innervation of the cerebral cortex has its main origin in the basal magnocellular nucleus, but fibers originating in the mesopontine cholinergic groups cannot be discounted (Foote and Morrison 1987; Reinoso-Suárez 1997). The cortical cholinergic innervation from the basal magnocellular nucleus is very well organized topographically (Figs. 2.41 and 3.5) (Reinoso-Suárez et al. 1982; Mesulam 1995). According to Mesulam (1995), all cytoarchitectonic regions and layers of the human cerebral cortex display dense cholinergic innervation. Cholinergic axon density is highest in the more superficial layers (layers I, II, and the upper parts of layer III) of the cerebral cortex. Major and statistically significant differences are also found in the overall density of cholinergic axons between the various cytoarchitectonic areas. The cholinergic innervation in primary sensory, unimodal, and heteromodal association areas is significantly lighter than that in paralimbic and limbic areas.

Discharge frequency fluctuates strongly in relation to behavioral tasks in basal forebrain cholinergic neurons in primates (Richardson and DeLong 1990). Single-unit studies in nonhuman primates have shown that the neurons of the basal forebrain are particularly sensitive to stimulus novelty and to the motivational relevance of the cues (Wilson and Rolls 1990). Some authors believe these cholinergic fibers provide information about the relationship between the internal and external environment, and all authors agree that acetylcholine is required for cortical activation and for correct functioning of the cerebral cortex in W and REM sleep (Reinoso-Suárez 1997). It is well known that cholinergic agonists produce cortical activation, whereas cholinergic receptor blockers, both nicotinic and muscarinic, decrease cortical activation. Semba (1999) writes that cholinergic axon terminals make symmetric synaptic contacts primarily with dendritic branches and less frequently, with spines and cell bodies of pyramidal and nonpyramidal excitatory neurons in rats, cats, and monkeys. However, Beaulieu and Somogyi (1991) showed that the cholinergic input is a dominant input for GABAergic cortical interneurons. Semba (1999) concludes that cholinergic basal forebrain afferents are targeted upon pyramidal cells as well as the excitatory and inhibitory interneurons in the neocortex, and, since cholinergic axon terminals in the cortex frequently do not form synaptic contacts, she proposes that, in addition to synaptic release, acetylcholine might be released extra-synaptically in a paracrine fashion. A similar proposal has been made regarding the monoaminergic terminals in the cerebral cortex. Thus, acetylcholine can stimulate cortical activation irrespective of behavioral arousal, motor activity and muscle tone, or REM sleep and muscle atonia. Henny and Jones (2008) showed that the major influence of the cholinergic input to the cortex would be largely through interneuronal innervations, and, possibly also, by diffusion of acetylcholine onto nearby pyramidal and other cell types. Perhaps, through contacts upon interneurons and nicotinic receptors, cholinergic cells can shape or pace activity in the cortex to stimulate theta rhythm during cortical activation, or else, through diffusion and muscarinic receptors, stimulate the slow depolarization in multiple cortical cells that promotes fast activity and prevents slow wave activity. One important mechanism of cholinergic action in the thalamus-cerebral cortex relationship is the potent muscarinic inhibition of GABA release from fast-spiking cells, the most prevalent interneuron subtype in the neocortex, to suppress thalamocortical feedforward inhibition. This is supplemented by the muscarinic-mediated depolarization of thalamic target-recipient excitatory neurons and the nicotinic enhancement of thalamic input onto these neurons that would promote thalamocortical excitation (Kruglikov and Rudy 2008).

In summary, acetylcholine can facilitate the transthalamic, and more precisely, the cortico-cortical and cortico-thalamo-cortical, processing of information in ways that may further modulate arousal and attention.

Serotonin. In all species, the density of serotonergic fibers is most uniform across the nuclei of the dorsal and ventral thalamus and epithalamus, and higher in the reticular and ventral lateral geniculate thalamic nuclei. The serotonergic thalamic inervation arises chiefly from the dorsal and central superior raphe nuclei (Azmitia and Segal 1978; Rodrigo-Angulo and Reinoso-Suárez 1982; Velayos and Reinoso-Suárez 1982; Jiménez-Castellanos and Reinoso-Suárez 1985; De Lima and Singer 1987; Westlund et al. 1990; Losier and Semba 1993).

Neocortical serotonergic innervation is very abundant in primates (Fig. 2.41). Serotonin release increases during awakening and is maintained during W. Serotonin has a tonic effect on cortical neuron activity. This effect is thought to occur through modulation of cortical neuronal responses in the phase changes in the sleep-wakening-W sequence and in the level of W (Jacobs et al. 1990). In the rhesus monkey at all ages, the cortical distribution patterns for serotonergic and catecholaminergic synthesis are the opposite of each other, indicating substantial interaction in the regulation of the two cortical systems (Goldman-Rakic and Brown 1982). What is more, development of monoaminergic storage capacity and synthetic processes continues over a period of months and years; this development is generally more rapid for serotonin than for catecholamines, and it varies greatly in different cytoarchitectonic regions of the cerebral cortex.

Adrenaline and noradrenaline. Most authors suggest that *adrenaline* is poorly distributed in the thalamus. Nevertheless, biochemical studies have shown the presence of adrenaline or its biosynthetic enzyme, phenylethanolamine-*N*-methyltransferase, in the rat and human thalamus. In rats, noradrenergic fibers have been described arising in the lower medulla and reaching the paraventricular nucleus and, very sparsely, paratenial nucleus (Otake and Ruggiero 1995). It has recently been demonstrated in the macaque monkey thalamus that the distribution of phenyletanolamine-*N*-methyltransferase-immunoreactive fibers is quite heterogeneous but principally restricted to the latter thalamic nuclei, or their portions, that are located in or close to the midline, with the highest density being found in the paraventricular, parafascicular, and mediodorsal nuclei (Rico and Cavada 1998). The entire paraventricular nucleus is densely innervated by adrenergic axons, while the densest innervation of the parafascicular nucleus is located in its medial part and the strongest mediodorsal nuclear immunolabelling is found in its most posterior and medial region. Moderate or low concentrations of phenyletanolamine-*N*-methyltransferase-immunopositive fibers are present in the paratenial nucleus, and all parts of the central nucleus, nucleus reuniens, central medial nucleus, centromedian nucleus, medial geniculate body, and medial pulvinar nucleus, while only scattered immunoreactive axons are found in other thalamic nuclei. Adrenaline may modulate a widespread number of cortical and subcortical brain structures through this thalamic innervation.

In monkeys, the highest thalamic *noradrenergic* fiber density is situated in the midline thalamic nuclei followed by the central lateral, centromedian, posterior, and suprageniculate nuclei. The anteroventral, anteromedial, paratenial, mediodorsal medial, ventral posteromedial, medial geniculate parvicellular, and pulvinar medial nuclei present moderate noradrenergic fiber density. Away from the dorsal thalamus there are a good numbers of noradrenergic fibers in the reticular thalamic nucleus, zona incerta, and lateral habenula (Rico and Cavada 1998). The density and extension of the noradrenergic fibers is lower in other mammals like the rat; however, in the latter, noradrenergic fibers are found in other nuclei such as the dorsal lateral geniculate and anterior nuclei. Noradrenergic terminals in the dorsal lateral geniculate nucleus are weak in the cat and absent in the monkey (Moore and Bloom 1979; Morrison and Foote 1986; De Lima and Singer 1987; Asanuma 1992; Rico and Cavada 1998; Bickford et al. 2000). The principal origin of the thalamic noradrenergic fibers appears to be the LCC, although fibers originating in other rhomboencephalic noradrenergic groups have been reported (De Lima and Singer 1987; Westlund et al. 1990; Losier and Semba 1993).

The LCC is the principal source of noradrenergic projections to the forebrain. Increasing tonic discharge by LC neurons elevates extracellular levels of noradrenaline in the cerebral cortex and thalamus. Functionally, substantial evidence indicates that the LC-noradrenaline projection system regulates behavioral state and state-dependent processing of sensory information. As we have described previously, tonic LC discharge is correlated with levels of arousal, and Devilbiss et al. (2006) have provided evidence linking LC neuronal discharge and noradrenaline efflux with LC-mediated modulation of single-neuron and neuronal ensemble representations of sensory stimuli in the ventral posteriomedial thalamus of awake rats. The output from the LC across a physiologic range modulates the responsiveness of single thalamic neurons to synaptic input. In Old and New

World monkeys, functionally related visual regions share common and distinguishable densities of noradrenergic innervation. Specifically, tecto-pulvinar-juxtastriate structures are more densely innervated than geniculo-striate and inferotemporal structures. These relationships suggest that, within the visual system, NA fibers preferentially innervate the regions involved in spatial analysis and visuomotor response rather than those involved in feature extraction and pattern analysis (Morrison and Foote 1986).

Also the widespread rich noradrenergic projections to the cerebral cortex originate in the LCC (Figs. 2.41 and 3.5). These projections are different in each species and in the different cortical areas, as we have already mentioned. The release of noradrenaline is higher in the cerebral cortex during W than in sleep (Foote and Morrison 1987). Levels of this neurotransmitter increase considerably on awakening and during attentional processes. The activation of the LC alone may be enough to induce electroencephalographic signs of cortical activation (Berridge and Foote 1991). Noradrenaline seems to highlight the evoked activity in relation with spontaneous activity. The LC is activated in the states of alertness and awakening, with noradrenaline increasing the selectivity and force of the cortical response to a stimulus. Noradrenaline seems to act on the thalamo-cortical and cortico-cortical circuit through changes localized in space within a short time (Foote and Morrison 1987). Experiments in rats have demonstrated that noradrenaline, and thus the activity of the LC, is necessary in the cerebral cortex for the induction of transcription factors, and this presence has potential effects on processes like learning and plasticity that occur during W (Cirelli et al. 1996). These results are consistent with previous findings in animals and humans: Coull et al. (1997) demonstrated that the significant attentional deficits induced by the alpha 2 adrenoceptor agonist clonidine are greater during states of relatively low arousal than during "cognitive" activation; these results suggest a neuroanatomical dissociation through thalamic and cortical noradrenergic modulation of arousal and attention.

In summary, and in agreement with Kayama and Koyama (2003), we may conclude that the noradrenergic projection from the LCC is an arousal system, since the noradrenergic neurons are active specifically during waking, and activation of the noradrenergic projection excites upper brain structures, in this case particularly the thalamus and cerebral cortex. The adrenergic and principally noradrenergic innervation of thalamic nuclei (especially midline and its neighbors) and their related cortical areas, such as limbic (specially orbitofrontal, cingular anterior, prefrontal medial ones), and association areas (both frontal and posterior association areas) show the cooperation of these thalamic and cortical regions in the functions of the awake brain.

Dopamine. Immunohistochemistry experiments on dopamine and dopamine transporter in the thalamus of macaque monkeys and humans have demonstrated that the dopamine innervation of the thalamus follows the same pattern in both species being densest in midline limbic nuclei, the mediodorsal and lateral posterior association nuclei, and in the ventral lateral and ventral anterior motor nuclei (García-Cabezas et al. 2007). In contrast, the scarce dopamine innervation of the rat thalamus is mainly located in the midline paraventricular nucleus, mediodorsal and ventral medial, and ventral lateral nuclei. In both primates and rats, dopamine innervation is consistent in the lateral habenula, reticular thalamic nucleus, and zona incerta (Fig. 3.10) (García-Cabezas et al. 2009). The densities of primate dopaminergic axons in the thalamus are as high as or higher than those in the most densely dopaminergically innervated cortical area. After ultrastructural analysis, the above authors conclude that in macaques, the thalamic GABAergic interneurons are a main postsynaptic target of DAT-ir axons; this suggests that the marked expansion of dopamine innervation in the primate in comparison to the rodent thalamus may be related to the presence of

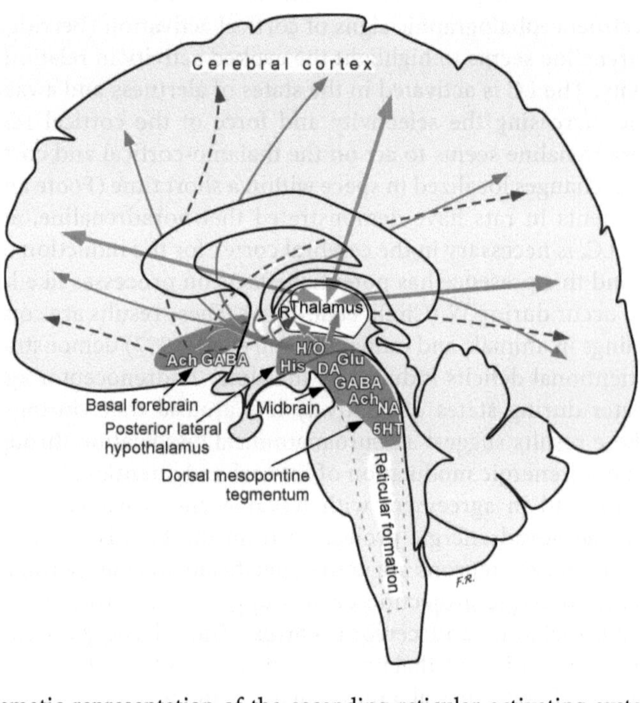

Fig. 3.10 Schematic representation of the ascending reticular activating system as understood in 2009. In a drawing of the medial view of a human right brain hemisphere, the main facts as known of the ascending reticular activating system are schematized as the neural base of W. The dorsal mesopontine and midbrain tegmenti, posterior lateral hypothalamus and basal forebrain structures, and also the neurotransmitters that project to thalamus and cerebral cortex directly wake the thalamus-cerebral cortex unit promoting fast bioelectrical activity in these structures and behavioral W. The *gray bidirectional arrows* that join the thalamus and cerebral cortex express the topographically close interconnections between both structures. *5-HT* serotonin neurons; *Ach* acetylcholine neurons; *DA* dopamine neurons; *GABA* GABA neurons; *Glu* glutamate neurons ; *His* histamine neurons; *H/O* hypocretin/orexin neurons; *R* reticular thalamic nucleus. Modified from Reinoso-Suárez (2005)

a larger interneuron population in primates. Also, the number of thalamic interneurons increases considerably in cats compared with rats. Consequently, the effect of dopaminergic innervation suppression may be different in cats and primates than in rats on functions like sleep and W. The origin of primate thalamic dopamine is multiple, and thus more complex, than in any other species studied so far: dopaminergic neurons of the hypothalamus, periaqueductal gray matter, ventral mesencephalon, and the lateral parabrachial nucleus project bilaterally to the monkey thalamus (Snchez-Gonzlez et al. 2005).

The dopaminergic fibers, which profusely innervate the cerebral cortex of primates, originate in the neurons of the ventral tegmental mesencephalic area (Figs. 2.41 and 3.5). These fibers are very abundant in the motor and association cortices and scanty in sensitive primary cortices. Dopamine transporter-immunoreactive (DAT-IR) axons are present throughout the primate cortical mantle, with substantial differences in density and laminar distribution across cytoarchitectonic areas. In particular, certain regions (e.g., posterior parietal cortex, dentate gyrus) not previously thought to receive a substantial dopamine input actually presented high densities of DAT-IR axons. The laminar distribution of the DAT-IR axons ranged from a restricted localization to layer I in lightly innervated regions to the presence of axons in all six cortical layers, with a particularly dense plexus in deep layer III in highly innervated regions (Lewis et al. 2001). Dopamine is released abundantly in active W compared with its steady release in the other phases of the SWC. It seems to be more effective in modulating cortico-cortical and cortico-thalamo-cortical than thalamo-cortical circuits, it is necessary for the organization of a suitable motor response, and it is believed to affect high-level integrative processes, like those occuring during working memory, more than analytic sensitive processes (see Foote and Morrison 1987; Sawaguchi and Goldman-Rakic 1994). It has been known for a long time that substances that increase the release of catecholamines, such as amphetamine, increase and prolong W, while substances that suppress catecholamine release diminish W (Reinoso-Suárez 1997).

As also occurs with other neurotransmitters, the patterns of regional and laminar distribution of dopaminergic axons differ significantly between primates and rodents. Thus, it appears that the major differences in dopamine innervation between the rodent and primate brain are in the cerebral cortex and dorsal thalamus (Fig. 3.10) (García-Cabezas et al. 2009). Compared with rodents, in primates the dopamine innervations in the cerebral cortex, in addition to presenting much larger reorganized terminal fields, a different phenotype for the localization of neuropeptides and very early prenatal development, are expanded and differently distributed: the innervation targets all cortical areas and is present in the deep and superficial layers with layer I being the most densely innervated in all areas (Berger et al. 1991; Williams and Goldman-Rakic 1993; Ciliax et al. 1999; Lewis et al. 2001). Thus, the wider and denser dopamine innervations in the primate dorsal thalamus and cerebral cortex as compared with the rodent may reflect the expansion of both structures in relation to encephalization (Berger and

Gaspar 1994; García-Cabezas et al. 2009); that is to say, in primates, there is a large growth in the extension of the higher order thalamic nuclei that is parallel to a proportional increase in the extension of the association areas of the cerebral cortex, and that both the thalamic nuclei and cortical areas are the most densely dopamine innervated brain regions. In addition, the primate thalamic nuclei have GABAergic interneurons that are absent in the rodent thalamus and in macaques; these thalamic interneurons appear to be the main postsynaptic target of the dopaminergic axons. All these observations imply a very different action of dopamine in the cerebral cortex and thalamus in rodents than in primates, and this may reflect a more significant involvement of dopamine in W promotion in the latter than in the former.

In primates, the dopamine activating effect is supported by the fact that dopamine forms inhibitory-type synapses with thalamic inhibitory interneurons (reticular and thalamic nuclei interneurons), so that dopaminergic excitation can disinhibit thalamic neurons projecting to the cortex, as happens in W (Fig. 2.33); moreover, dopaminergic fibers make wide connections, in a pattern that may permit greater neurotransmitter diffusion, with cellular elements that are closely related with SWC phenomenology and located in layer I of the cerebral cortex (Lewis et al. 2001; Chap. 2 of this publication). This happens mainly in the prefrontal, motor and posterior association (especially posterior parietal) cortices, and corresponding related nuclei that have a significant role in active W and attention functions.

GABA. Thalamic GABAergic fibers and terminals, in addition to those originating in the intrinsic thalamic interneurons and neurons of the thalamic reticular nucleus, have been reported. Most come from the medial globus pallidus and the reticular substantia nigra. Basal forebrain GABAergic neurons and GABAergic fibers from the zona incerta were found to selectively innervate relay cells located mainly in higher-order thalamic nuclei. These large GABA-immunoreactive terminals established multiple contacts preferentially on the proximal dendrites of relay cells via symmetrical synapses with multiple release sites (Barthó et al. 2002). It has been demonstrated that the ascending inputs from cholinergic brainstem nuclei, which would operate during W, suppress the zona incerta GABAergic inhibitory mechanism of the thalamic projection neurons during NREM sleep, representing another path to modulate the flow of information through thalamic nuclei by means of the cholinergic system (Masri et al. 2006). This may occur chiefly in the higher-order thalamic nuclei projecting to the association cortices; the proximally located multiple active zones of ZI terminals indicate a powerful influence on the firing properties of thalamic neurons, one that is conveyed to multiple cortical areas chiefly via the M-type cells with widespread projections to neocortex, mainly to layer I (Barthó et al. 2002; Rubio-Garrido et al. 2009). The zona incerta also projects to the supragranular layers of the neocortex, especially layer I (Nicolelis et al. 1992).

It has been supposed that one-third of the basal forebrain cortical projections in cats are GABAergic, another third cholinergic and the remaining third would be noncholinergic-non-GABAergic (Fisher et al. 1988). Similar proportions have been reported more recently in rats. The main target of GABAergic basal forebrain projections to the cerebral cortex, which made symmetrical synapses, appears to be GABAergic interneurons. Consequently, the result of the activation in GABAergic basal forebrain neurons might be a transient disinhibition of pyramidal neurons (Semba 1999).

Barthó et al. (2002) have shown that cortical terminals establish asymmetrical synapses on zona incerta cells with very large active zones. These observations suggest efficient integration of widespread cortical signals by single zona incerta neurons and a strong cortical drive. Those authors proposed that the efferent GABAergic signal of zona incerta neurons as patterned by cortical activity could play a critical role in thalamocortical rhythms.

Histamine. Sallmen et al. (1999) described low histaminergic fiber density in the anteroventral, ventrolateral, dorsal, ventroposterior, posterior, pretectal, parafascicular, and lateral geniculate nuclei in the ground squirrel thalamus (Fig. 3.4). Moderate fiber density was found in the paratenial, anteromedial, periventricular, mediodorsal, habenular, and medial geniculate nuclei. The reuniens thalamic nucleus displayed high fiber density. H1 and H3 receptor mRNAs were relatively enriched in the anterior, medial, and part of the lateral nuclei regions of normal human thalamus; however, the expression level was much lower in the ventral and posterior parts, as well as, in the reticular nucleus (Jin et al. 2002). The intensity of H2 receptor mRNA expression was generally very low. H1 receptor binding was mainly detected in the mediodorsal, ventroposterolateral, and pulvinar nuclei. H3 receptor binding was detected mainly in the dorsal thalamus and was predominant in the periventricular, mediodorsal, and posterior regions. Very high or high histaminergic fiber densities were observed in the midline nuclear region and other nuclei next to the third ventricle, ventroposterior lateral nucleus, and medial geniculate nucleus. Fiber density was very heterogeneous in most of the core structures of the thalamus. The results in human thalamus suggest that histamine regulates some functions through multiple receptors, and despite the scantiness of histaminergic nerve fibers, histamine receptors expressed in thalamic nuclei. The presence of H3 receptors on thalamic nerve terminals has been recently described, and their activation modulates presynaptic glutamatergic, but not GABAergic, neurotransmission (Garduño-Torres et al. 2007). Application of histamine to the perigeniculate nucleus blocked the generation of spindle waves, indicating that increased activity in the tuberomammillary histaminergic system may play a functional role in dampening thalamic oscillations in the transition from sleep to arousal (Lee et al. 2004).

Hypothalamic histaminergic neurons innervate the cerebral cortex. Histaminergic agonists increase W and cortical activation and antihistaminic drugs cause drowsiness. Neocortical and allocortical areas of the ground squirrel cerebral cortex contained moderately dense histaminergic fiber networks (Fig. 3.4) (Sallmen et al. 1999). A well-organized network of varicose fibers was revealed throughout the frontal and temporal cortex of adult humans with specific antisera against histamine. The densest fiber network was seen in lamina I, where varicose fibers ran parallel to the overlying pia mater. Jin and Panula (2005) have employed in situ hybridization and receptor binding autoradiography to map and quantify the mRNA expression and receptor binding of three of the four histamine receptors (H1, H2, H3). mRNA expression and receptor binding for these three receptors displayed characteristic laminar distribution patterns. Both H1 and H3 receptor mRNAs were mainly expressed in the deeper layers (H1 in layers V and VI; H3 in layer V), where most of the corticothalamic (and other subcortical) projections originate, whereas H2 receptor mRNA was primarily expressed in superficial layer II. Higher densities of H1 and H3 receptor radioligand binding sites were seen in middle layers III and IV, which receive abundant thalamic inputs and are where some of the apical dendrites of the deep-layer pyramidal neurons terminate, whereas H2 receptor radioligand binding site density was higher in superficial layers I-III. Those authors suggest that histamine regulates both cortico-cortical and thalamo-cortical circuits.

A histaminergic influence on neocortical synaptic plasticity has recently been demonstrated in vivo and it has also been shown that cortical histaminergic activation lowers the induction threshold as well as increases the degree of plasticity in the mature thalamocortical visual system (Kuo and Dringenberg 2008). Examining regional cerebral blood flow (rCBF) responses using positron emission tomography (PET) and H2O during a simulated car-driving task following oral administration of the antihistamine D-chlorpheniramine, Tashiro et al. (2008) found that the parietal, temporal and visual cortices, and the cerebellum showed decreased brain responses. Meanwhile, the rCBF responses were increased in the orbitofrontal cortex and cerebellar vermis. The authors suggest that D-chlorpheniramine tends to suppress visuo-spatial cognition and visuomotor coordinating functions rather than attention and motor functions during car driving.

The concurrent action of histamine in the basal forebrain cholinergic complex, the thalamus, and the cerebral cortex probably allows histamine to effectively modulate cortical activation and excitability through different behavioral states (Dringenberg and Kuo 2003). In summary, these thalamic and cortical results are coherent with the suggestion that the hypothalamic histaminergic neurons are one of the main structures responsible for cortical activation during W (Lin 1994).

Hypocretin/Orexin. The density of Hcrt/Orx-IR fibers in the rat thalamus is high in the central gray and paraventricular, central medial, subparafascicular, and lateral habenular nuclei; lower in the lateral habenular, reuniens, rhomboid, and parafascicular nuclei; poor in the reticular, ventral lateral geniculate, and laterodorsal nuclei; and very poor in mediodorsal, ventral lateral, ventral posterior, lateral, and medial geniculate nuclei (Fig. 3.7) (Peyron et al. 1998). The human thalamus also presents a dense orexinergic innervation of the midline nuclei (Moore et al. 2001). These nuclei project to limbic cortical areas such as the orbitofrontal, cingular, prefrontal medial, and anterolateral as well as frontal, parietal, and temporal association cortices.

Hcrts/Orxs may selectively excite nonspecific thalamocortical projection neurons (M-type) in nuclei close to the midline – the centromedial and rhomboid nuclei – (Bayer et al. 2002b), ones that project to the limbic and, particularly, the prefrontal cortex (Rubio-Garrido et al. 2009). Bayer et al. (2002b) thus concluded that Hcrts/Orxs could act in the thalamus to promote W by exciting neurons of the nonspecific thalamocortical projection system (M-type), which stimulates and mantains cortical activation through its projections to wide territories of the cerebral cortex. In fact, nonspecific projection neurons are unusual in that they have burst-pattern firing during W, when Hcrts/Orxs levels are elevated, instead of bursting during NREM sleep (Steriade et al. 1993). These results fit with those of functional imaging studies showing that this thalamic system presents higher levels of activation during arousal and attention, results that confirm the role of nonspecific projection neurons in arousal and W (Kinomura et al. 1996). As we have described in the previous chapter, these neurons project to layer 1 apical dentrites of pyramildal cells whose somas are situated in cortical layers II, III, and V. In fact, the Hcrts/Orxs neurons can have both a direct and indirect influence upon thalamocortical transmission: direct through the projections to the thalamus and indirect through the projections to brainstem, hypothalamic, and basal forebrain waking neurons that, in turn, project to the thalamus.

These results and considerations are consistent with those reported more recently by Govindaiah and Cox (2006), who examined the actions of Hcrts/Orxs on thalamic neurons using an in vitro rat thalamic slice preparation. The orexins (orexin-A and orexin-B) produced distinct actions within different thalamic nuclei. Orexin-B strongly depolarized 71% of the central lateral nucleus neurons, 10% of the parafascicular nucleus neurons, and 21% of the mediodorsal thalamic nucleus neurons tested. Overall, Orexin-B was found to be more potent than Orexin-A. Orexin-A produced a small depolarization in 28% of the neurons in the thalamic reticular nucleus. Neither Orexin-A nor Orexin-B had any effect on neurons in the lateral posterior, laterodorsal, posterior, ventrobasal, and lateral geniculate thalamic nuclei. The depolarizing actions of orexins were sufficient to alter the firing mode of these neurons from a sleep-like burst to a W-like tonic-firing mode. The excitatory actions of orexin-B result from a decrease in the apparent potassium current leak (K-leak). The actions of Orexin-B were blocked by the classical neurotransmitter dopamine, indicating that Hcrt/Orx may share similar ionic mechanisms. Probably, the waking action of the two neurotransmitters on the thalamus is different and stronger in primates than in rodents.

In the rat, the Hcrt/Orx fibers innervating the cerebral cortex are mainly long and thin with varicosities and show medium density in deep layers, lighter in layer I–III, and very light in layer III (Peyron et al. 1998). In humans, the Hcrt/Orx neurons lightly innervate all areas of cerebral cortex studied with a variable pattern that tends to be denser in the innervations of the association cortex than in the primary motor or sensory cortices (Moore et al. 2001). Independently of direct cortical excitation of the Hcrt/Orx neurons, as occurs in the thalamus, these neurons may drive the wake-active neurons in the bainstem, hypothalamus, and, as has recently been demonstrated by Murillo-Rodriguez et al. (2008), in the basal forebrain, thus indirectly arousing the cerebral cortex; therefore, a weak Hcrt/Orx flow into the wake-promoting neurons in the basal forebrain, such as after Hcrt/Orx neuron lesion, would not allow the cortex to wake up adequately. Also the glutamatergic paraventricular thalamic nucleus neurons, which receive one of the densest Hcrts/Orxs thalamic innervations, project, in addition to other cortical areas, to the medial prefrontal cortex. These cortically projecting neurons are depolarized and excited by Hcrt-2. The robust excitation evoked by Hcrt-2 in cortically projecting glutamate paraventricular neurons could generate substantial excitation in multiple layers of the medial prefrontal cortex, adding to the more selective direct excitatory actions of Hcrts/Orxs in this cortex and potentially increasing cortical arousal and attention through this indirect pathway (Huang et al. 2006).

However, not only does Hcrts/Orxs excite nonspecific thalamic projection neurons, as described in previous paragraphs, Hcrt/Orx projections also stimulate the terminal projections of these thalamic neurons in the cerebral cortex (Lambe and Aghajanian 2003; Lambe et al. 2007). It seems that the Hcrt/Orx cortical projection potently increased synaptic glutamate release onto these thalamocortical terminals. This effect of Hcrt/Orx occurred preferentially in the rat medial prefrontal cortex, a region that has been shown to receive a greater density of Hcrt/Orx projections than other cortical regions in rats. In consequence, Hcrt/Orx stimulation appears to depolarize thalamocortical terminals to reach the spiking threshold. In summary, there is evidence that Hcrt/Orx excites thalamocortical synapses in prefrontal cortex. The mechanism for this action appears to be that Hcrt/Orx excites, generally at the level of layer I, single identified thalamocortical M-type neuron synapses on apical dendrites of layer V neurons in the prefrontal cortex without producing direct postsynaptic effects in layer II/III or layer V pyramidal neurons. By inducing spikes in thalamocortical terminals, Hcrt/Orx release in the prefrontal cortex has the potential to affect alertness and attention in vivo (Lambe et al. 2007). These authors also found that nicotine excites high-affinity nicotinic acetylcholine receptors on the same thalamocortical terminals that are activated by Hcrt/Orx and results in a large increase in glutamate release onto prefrontal layer V pyramidal neurons (Lambe et al. 2005). Similar events may occur in other cortical areas.

The strongest evidence from research data on Hcrts/Orxs regards their involvement in the integration and stabilization of arousal networks; de Lecea and Sutcliffe (2005) suggest that the hypocretinergic system integrates homeostatic, metabolic and limbic information to provide a coherent output that results in stability among the states of vigilance.

Other neurotransmitters at cellular level. The thalamic and cortical projections from the pontine and mesencephalic tegmentum (including some that originate in the peribrachial region), which are not monoaminergic or cholinergic, may be largely *glutamatergic* since glutamatergic neurons are abundant in those areas; in fact, acetylcholine even colocalizes with glutamate in some neurons. Additionally, many hypothalamic and basal forebrain connections to thalamus and cerebral cortex are glutamatergic. Finally, the thalamo-cortical and cortico-thalamic connections use glutamate as a neurotransmitter, as do the cortical projections to the subcortical structures related with W organization.

Many other neurotransmitters take part in the neuronal network that supports the organization of W; some of them have been addressed in previous sections of this revision. In general, agonists of the neurotransmitters that activate waking structures increase W and cortical activation while the antagonists cause sleepiness and cortical synchronization.

There are data that show a correlation of unit activity in thalamic and cortical neurons with the process of release of neurotransmitters involved in each phase of the SWC. In the thalamus, nonrelease of neurotransmitters implicated in the "ascending reticular activating system" during light and deep NREM, by both projecting thalamocortical neurons as well as reticular thalamic neurons, allows the triggering of circuits with intrinsic oscillations that can be expressed in sleep spindles or slow waves, the latter with a cortical pacemaker (Steriade et al. 1991; Sillito et al. 1994). These slow oscillations of the thalamic cells can be facilitated by cooperation from the thalamic inhibitory interneurons when these cease to be inhibited by acetylcholine (McCormick 1992). In the thalamus, the activation of the histaminergic, noradrenergic, muscarinic-cholinergic, and metabotropic-glutamate receptors suppresses the potassium current and thereby induces slow depolarization (McCormick and Pape 1988; Curró-Dossi et al. 1991; McCormick and Williamson 1991; Jin et al. 2002). Consequently, the increased release of monoamines, histamine, acetylcholine, glutamic acid, and Hcrt/Orx occuring when moving into the W phase inactivates the low-threshold Ca^{2+} currents. Loss of these currents arrests the slow-bursting oscillation of the thalamo-cortical and reticular thalamic nucleus neurons and sets off a tonic firing pattern with single spikes in these neurons. Both thalamic inhibitory interneuron inhibition and thalamocortical C- and M-type neuron excitation contribute to this action (McCormick and Bal 1994). Under these conditions, awakening and active W, information reaching the thalamus can be transmitted accurately to the cerebral cortex by the generation of action potentials in the C-type neurons; the excitation in W and the single spike firing by the M-type neurons produce a widespread excitation at cerebral cortex layer I that contributes to cortical electroencephalographic activation.

In experiments at the cellular level it has also been demonstrated that the increased release of monoamines, histamine, acetylcholine, and glutamate in the cerebral cortex during arousal and active W would suppress the slow activity of the cortical neurons and switch them to a tonic firing with simple spikes (McCormick 1992); Hcrt/Orx also contribute to this excitation expressed as cortical electroencephalographic activation. Thus, the cerebral cortex, the substrate for perception and complex cognitive processes, is ready to process the abundant information that it receives from the thalamus and other cortical areas during W.

3.2.3
The "Thalamo-Cortical Unit"

As we have commented in Chap. 2, the thalamus and the cerebral cortex work as an indivisible unit in brain functions. Most of the brainstem, hypothalamic, and basal forebrain "ascending reticular activating system" acts directly on both structures during W, as we have seen in the previous section (Figs. 2.32, 2.33, 2.43, 3.5, 3.9, and 3.11). The thalamo-cortical relations change considerably from rodents to carnivores and primates. In the last two groups, both the extension of the higher-order thalamic nuclei and, proportionally, the extension of the association areas of the cerebral cortex grow; also the carnivore and primate thalamus have GABAergic interneurons that are absent from the rodent thalamus. In addition, the rodent thalamus is mainly composed of first-order thalamic relay nuclei while association cortices are scant. The first-order thalamic relay nuclei usually transmit information from a subcortical source to a primary sensory or motor cortical area. In contrast, the higher order thamic nuclei act as intermediate links to transfer information from layer V of an association cortical area to other association cortical areas. In carnivores and primates, most of the thalamus seems to be comprised of higher-order nuclei that are especially important for cortico-cortical communication (Reinoso-Suárez 1984).

In any case, the cerebral cortex is the main afferent and efferent structure for the thalamus, and, in turn, the thalamus is the main subcortical afferent and efferent structure for the cerebral cortex (Figs. 2.32, 2.33, and 3.5). In both cases, these connections use, as we have described in Chap. 2, two types of projection neurons: the thalamus uses C- and M-type thalamo-cortical projection neurons and the cerebral cortex uses projection neurons from layer VI and layer V. All four types of neurons are active during W because the brainstem, hypothalamic and basal forebrain waking neurotransmitters, and neuromodulators suppress slow activity in these neurons, which then switch to firing tonically with single spikes (McCormick 1992; McCormick and Bal 1994). In this functional state, during awakening and active W, the information that arrives at the thalamus can be transmitted accurately to specific areas of the cerebral cortex by the generation of action potentials in C-type neurons; excitation in W and firing with simple spikes by the M-type neurons may produce a direct widespread excitation at the level of cerebral cortex layer I that contributes to cortical electroencephalographic activation (Fig. 2.33). In some of these cases, the M-type cells project to a limited single area of cerebral cortex layer I or to a few adjacent cortical areas (Rubio-Garrido et al. 2009). In every case, interactions between M-type and C-type thalamo-cortical input terminals may occur, and if they do, they improve pyramidal cell firing and synaptic plasticity (Spruston 2008; Rubio-Garrido et al. 2009). It is important to emphasize that these mechanisms are very complex and that the waking neurotransmitters operate not only at the thalamic and cortical neuronal level but also, as we have described for Hcrt/Orex, by exciting thalamocortical

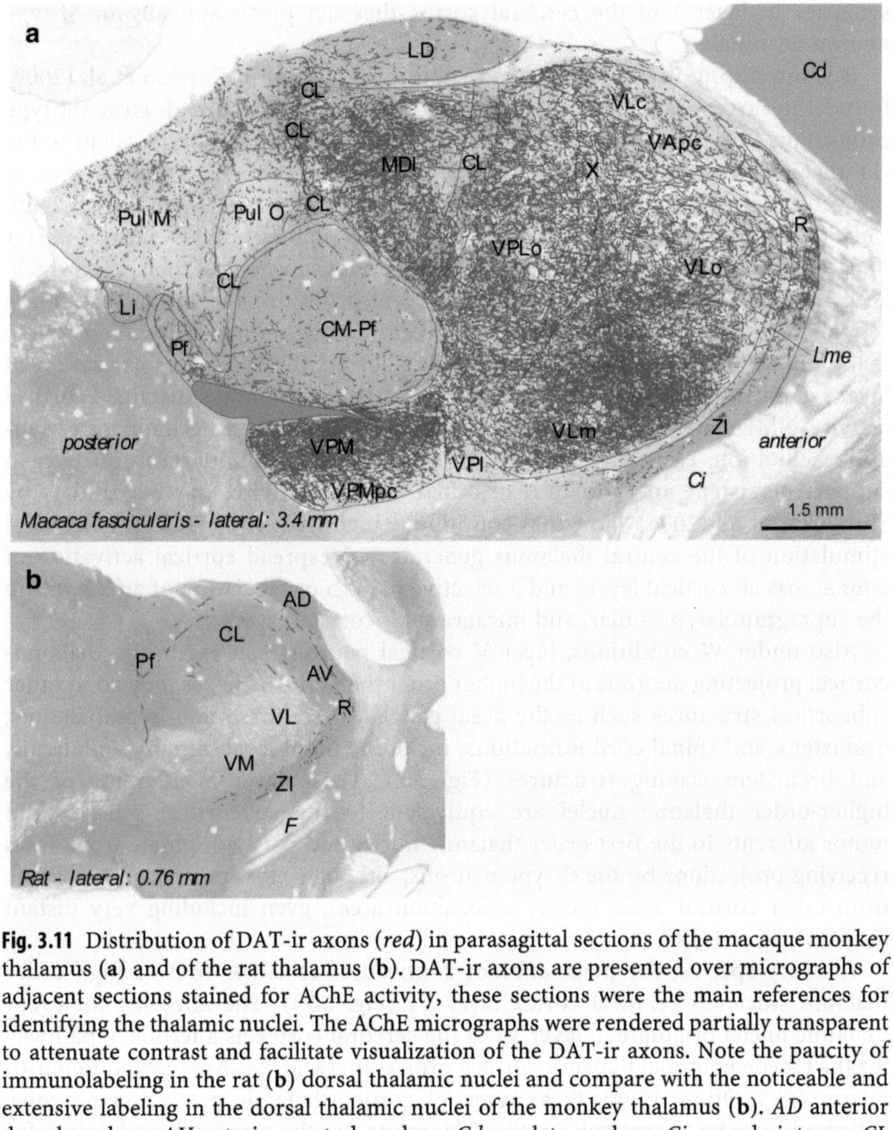

Fig. 3.11 Distribution of DAT-ir axons (*red*) in parasagittal sections of the macaque monkey thalamus (**a**) and of the rat thalamus (**b**). DAT-ir axons are presented over micrographs of adjacent sections stained for AChE activity, these sections were the main references for identifying the thalamic nuclei. The AChE micrographs were rendered partially transparent to attenuate contrast and facilitate visualization of the DAT-ir axons. Note the paucity of immunolabeling in the rat (**b**) dorsal thalamic nuclei and compare with the noticeable and extensive labeling in the dorsal thalamic nuclei of the monkey thalamus (**b**). *AD* anterior dorsal nucleus; *AV* anterior ventral nucleus; *Cd* caudate nucleus; *Ci* capsula interna; *CL* central lateral nucleus; *CM-Pf* centromedian-parafascicular complex; *F* fornix; *LD* lateral dorsal nucleus; *Li* limitans nucleus; *Lme* lamina medular externa; *MDl* mediodorsal nucleus-lateral sector; *Pf* parafascicular nucleus; *R* reticular nucleus; *VApc* ventral anterior nucleusparvocellular part; *VL* ventral lateral nucleus; *VLc* ventral lateral nucleus-caudal part; *VLm* ventral lateral nucleus-medial part; *VLo* ventral lateral nucleus oral part; *VM* ventral medial nucleus; *VPI* ventral posterior inferior nucleus; *VPLo* ventral posterior lateral nucleus-oral part; *VPM* ventral posterior medial nucleus; *VPMpc* ventral posterior medial nucleus parvocellular part; *X* ventral lateral nucleus-area X; *ZI* zona incerta. Modified from García-Cabezas et al. (2009)

synapses in layer I of the cerebral cortex they act presynaptically in M-type neuron terminals.

It is interesting that the thalamic region in which Rubio-Garrido et al. (2009) found the largest number of M-type neurons with widespread axon-subtype projections to layer I in multiple distant cortical areas in rats is equivalent to the cat thalamic paralaminar region where we have described thalamic neurons projecting to layer I of the frontal and parietal cortices (see Chap. 2; Fig. 2.39). This region includes the ventral medial, ventral lateral, and ventral anterior thalamic nuclei and may represent, in these species, the central thalamus, which, according to Schiff (2008), receives convergent projections from the "ascending reticular activating system" and the frontal cortex, thus emphasizing their role in maintaining organized behavior during W. For Schiff (2008) depolarization of layer I dendrites by M-type neuron thalamic afferents promotes sustained cortical activity, NMDA-mediated long-term potentiation, and other mechanisms of synaptic facilitation. Thus, the M-type neuron may provide an important substrate to support persistent and widely distributed cortical activity. In vivo studies by Shirvalkar et al. (2006) show that continuous high frequency (100 Hz) electrical stimulation of the central thalamus generates widespread cortical activation of *c-fos* across all cortical layers and a selective pattern of regulation of *zif268* within the supragranular, granular, and infragranular cortical layers.

Also under W conditions, layer V cortical neurons can excite the thalamo-cortical projecting neurons of the higher order thalamic nuclei, as they do in other subcortical structures such as the basal ganglia, basal forebrain, hypothalamus, brainstem, and spinal cord formations, including basal forebrain, hypothalamic, and brainstem waking structures (Fig. 3.5). These layer V afferents of the higher-order thalamic nuclei are equivalent to the subcortical sensory and motor afferents to the first-order thalamic nuclei and partly originate from areas receiving projections by the C-type neurons, although most projections originate from other cortical areas mostly association areas, even including very distant cortices (Fig. 2.33).

In every species, the main afferent contributor to either first or high order thalamic nuclei is cerebral cortex layer VI (Fig. 2.33). The afferents for every thalamic nuclei originate in layer VI of the cerebral cortex as a feedback pathway for the C-type neuronal thalamo-cortical projection (Rodrigo-Angulo and Reinoso-Suárez 1995; Ojima 1994); these layer VI cortico-thalamic projection neurons, activated in W by the reticular activating inputs, are the principal modulators for reticular and projection thalamic neuron activity. The layer VI cortico- thalamic projection would decode the appropiate nature and format of the information that must reach this cerebral cortex from the thalamus, and this of course done jointly with the brainstem, hypothalamic, and basal forebrain waking influences. Also it must not be forgotten that asynchronous convergence during W by afferents (from the periphery – in first order nuclei – or cerebral cortex layer V – in the higher order nuclei – as well as the reticular activating system) and projections from cerebral cortex layer VI can dynamically alter the physiological properties of thalamo-cortical projection neurons (Figs. 2.33 and 2.41) (Sherman 2001; Nicolelis and Fanselow 2002).

However, these actions exerted by brainstem, hypothalamic, and basal forebrain waking influences are cell-specific in the thalamus and cerebral cortex. As an example, Devilbiss and Waterhouse (2004) demonstrated that LC output can produce several different neuromodulatory actions simultaneously on many single neurons within the ventral posterior thalamic nucleus and somatosensory cortex of the rat and that these actions are dependent on LC output levels in a cell-specific manner. Differences in modulatory actions between the thalamus and cortex suggest that LC output has unique effects on signal processing operations at each level of an ascending sensory network. The observed effects on probability of neuronal responsiveness and response timing suggest that sensory signal processing is continually altered over the range of tonic LC discharge frequencies. They hypothesize that higher frequencies of LC output, as they occur during active W, regulate target neurons across a wide dynamic range of responsiveness to maintain the greatest flexibility in processing environmental signals that are relevant to ongoing behaviors. These findings suggest that, over the range of tonic LC output frequencies, target cell networks continuously self-organize to optimize performance in ongoing behavioral tasks. The reciprocal influences of the wakening cortical and thalamic neurons take an important role in this optimal performance.

In summary, the "thalamo-cortical unit" is awakened by the "ascending reticular activating system" with its brainstem, hypothalamic, and basal forebrain components; however, during W the complex interrelated structures of this unit modulate, organize, and emphasize the mechanisms for the adequate and appropriate activity of the different thalamic and cortical formations during W, thus allowing the organization of cognitive processes and the performance of adequate behavioral responses. The collective activity of these ascending pathways provides the physiological condition supporting the precise pathways that interconnect cortex and thalamus, and these, with other brain structures, would be the machinery for complex behaviors related to cognition and behavior during W. The "thalamo-cortical unit" efferents for this purpose are located in cerebral cortex layer V, the pyramidal neurons which act on subcortical structures such as the basal ganglia, hypothalamus, brainstem, cerebellum, and spinal cord. Achieving an adequate W state in this situation requires adjusting the activation pattern across the brain structures by joint tonic levels of the "reticular ascending activating system" that controls levels of thalamic and cortical activity and top-down signals from the "thalamo-cortical unit" to the brainstem; the hypothalamic and basal forebrain neurons are the origin of the system (Fig. 3.5). The top-down signals originate in layer V pyramidal neurons of the association cortices, mainly frontal, cingular, and parietal regions, involved in monitoring demands for attention and adjustment of vigilance level or alertness. Now, during W, we are aware, "we feel, we think, and we work in full knowledge of what we do" and "fully use our senses and faculties."

There is still another efferent in the thalamus-cerebral cortex unit and that is the thalamo-striatal projection. Thalamo-striatal excitatory input can act as a primary driver and, along with descending cortico-striatal inputs, may organize very long-range functional connections in the brain during W. In addition, the amygdaline complex receives inputs from both the thalamus and the cerebral cortex.

3.3
Final Commentary

The "ascending reticular activating system" maintains the "thalamo-cerebral cortex unit" awake (Figs. 3.5 and 3.9). This system originates from brainstem, hypothalamic, and basal forebrain cell groups with known neurotransmitters. Neurons in each of these groups have the property of being most active during W. The release of several neurotransmitters (acetylcholine, noradrenaline, serotonin, GABA, glutamate, histamine, Hcrt/Orex, etc.) by the "ascending reticular activating system" depolarizes thalamocortical neurons and excites cortical pyramidal cells, thereby suppressing the generation of sleep rhythms and promoting a W state that is favorable for sensory processing and cognition (McCormick and Bal 1997; Reinoso-Suárez 1997).

This W state is expressed in the cerebral cortex with an activated (desynchronized) EEG with fast gamma activity (30–60 Hz) during vivid attention or motor processing. EEG activation (including gamma activity) is also present during the REM sleep state, perhaps because in both active W and REM sleep, as has been demonstrated in single-unit recording studies, the brainstem and basal forebrain cholinergic neurons discharge at high rates in association with cortical gamma activity (Alonso 1998; Kayama and Koyama 2003; Lee et al. 2005b). However, in REM sleep, in addition to other neurobiological phenomena, the enhanced thalamic and cortical activity depends chiefly on a single neurotransmitter, acetylcholine (and perhaps glutamate), whereas in W, this enhanced activity results from the harmonized cooperation of a variety of neurotransmitters (acetylcholine, noradrenaline, serotonin, GABA, glutamate, histamine, Hcrt/Orex, etc.) acting on the "thalamo-cortical unit." Although some cholinergic neurons do not discharge in both W and REM sleep, the ones that do participate in both states are modulated in W by other neurotransmitters and may discharge at a different rate in each state (el Mansari et al. 1989; Kayama et al. 1992; Kayama and Koyama 2003).

Therefore, in order for the cortical and thalamic activation of W to occur with all its characteristic plastic nuances in its different behavioral circumstances, there must be synergy among several of the neurotransmitters systems in the "ascending reticular activating system." However, there are data showing that the different neurotransmitters use different strategies to achieve the same result. Thus, recently, Kruglikov and Rudy (2008) showed that the inhibition of GABA release from fast-spiking cortical cells produced by serotonin, muscarine, adenosine, and baclofen modulates the dynamics of thalamocortical activation. The modulations characterized in their study appeared to be mediated by two different mechanisms, although all four modulators resulted in an inhibition of GABA release. Those authors showed that the effect of serotonin had a relatively slow onset and involved the activation of protein kinases by Gq/11 proteins coupled to 5-HT2 receptors. On the other hand, the modulation of GABA release produced by muscarine, adenosine, and baclofen displayed a fast onset and was independent of protein kinase activation, thereby implicating the involvement of membrane-delimited mechanisms. Nonetheless, we know little, even today, of the interaction among the different neurotransmitters, although we do have data showing that every system of neurotransmitters that helps to awaken and maintain W also modifies the activity of the other neurotransmitters through heteroceptive actions (Marrocco et al. 1994; Reinoso-Suárez 1997). Each of the "ascending reticular activating system" neurotransmitters that we have considered and which activate or modulate the thalamus and cerebral cortex during W can exert a powerful influence on information processing and the attentional, emotional, motivational, and arousal states.

The awake "thalamo-cerebral cortex unit" controls and adjusts the activation pattern through a top-down action on the brainstem, hypothalamic, and basal forebrain cellular groups, which are the origin for the "ascending reticular activating system," and this top-down action is necessary and essential for an adequate W state (Fig. 3.5).

References

Abbott A (2005) Neuroscience: while you were sleeping. Nature 437:1220–1222

Adamantidis AR, Zhang F, Aravanis AM, Deisseroth K, de Lecea L (2007) Neural substrates of awakening probed with optogenetic control of hypocretin neurons. Nature 450:420–424

Adler CH (2005) Nonmotor complications in Parkinson's disease. Mov Disord 20:S23–S29

Alonso A (1998) Intrinsic electroresponsiviness of basal forebrain cholinergic and non cholinergic neurons. In: Lydic R, Baghdoyan HA (eds) Handbook of behavioural state control. Cellular and molecular mechanisms. CRC Press, Boca Raton, London, New York, pp 297–309

Arbuthnott GW, MacLeod NK, Maxwell DJ, Wright AK (1990) Distribution and synaptic contacts of the cortical terminals arising from neurons in the rat ventromedial thalamic nucleus. Neuroscience 38:47–60

Arpa J, De Andres I (1993) Re-examination of the effects of raphe lesions on the sleep/wakefulness cycle states in cats. J Sleep Res 2:96–102

Asanuma C (1992) Noradrenergic innervation of the thalamic reticular nucleus: a light and electron microscopic immunohistochemical study in rats. J Comp Neurol 319:299–311

Asanuma C (1997) Distribution of neuromodulatory inputs in the reticular and dorsal thalamic nuclei. In: Steriade M, Jones EG, McCormick DA (eds) Thalamus, experimental and clinical aspects, vol II. Elsevier, Amsterdam, pp 93–153

Aserinsky E, Kleitman N (1953) Regularly occurring periods of eye motility, and concomitant phenomena during sleep. Science 118:273–274

Aston-Jones G, Bloom FE (1981) Activity of norepinephrine-containing locus coeruleus neurons in behaving rats anticipates fluctuations in the sleep-waking cycle. J Neurosci 1:876–886

Aston-Jones G, Chiang C, Alexinsky T (1991) Discharge of noradrenergic locus coeruleus neurons in behaving rats and monkeys suggests a role in vigilance. Prog Brain Res 88:501–520

Avendaño C, Llamas A (1983) Thalamic and nonthalamic direct subcortical projections to association areas of the cat's cerebral association cortex. In: Reinoso-Surez F, Ajmone Marsan C (eds) Cortical integration: basic, archicortical and cortical associa- tion levels of neural integration, vol 11, IBRO monograph series. Raven Press, New York, pp 195–222

Avendaño C, Stepniewska I, Rausell E, Reinoso-Surez F (1990) Segregation and heterogeneity of thalamic cell populations projecting to superficial layers of the posterior parietal cortex: a retrograde tracing study in the cat and monkey. Neuroscience 39:547–559

Azmitia EC, Segal M (1978) An autoradiographic analysis of the differential ascending projections of the dorsal and median raphe nuclei in the rat. J Comp Neurol 179:641–667

Baghdoyan HA, Rodrigo-Angulo ML, McCarley RW, Hobson JA (1984) Site-specific enhancement and suppression of desynchronized sleep signs following cholinergic stimulation of three brainstem regions. Brain Res 306:39–52

Barthó P, Freund TF, Acsdy L (2002) Selective GABAergic innervation of thalamic nuclei from zona incerta. Eur J Neurosci 16:999–1014

Batini C, Moruzzi G, Palestini M, Rossi GF, Zanchetti A (1958) Presistent patterns of wakefulness in the pretrigeminal midpontine preparation. Science 128:30–32

Batini C, Moruzzi G, Palestini M, Rossi GF, Zanchetti A (1959) Effects of complete pontine transections on the sleep-wakefulness rithms: the midpontine pretrigeminal preparation. Arch Ital Biol 97:1–12

Bayer L, Eggermann E, Saint-Mleux B, Machard D, Jones BE, Mühlethaler M, Serafin M (200a) Selective action of orexin (hypocretin) on nonspecific thalamocortical projection neurons. J Neurosci 22:7835–7839

Bayer L, Mairet-Coello G, Risold PY, Griffond B (200b) Orexin/hypocretin neurons: chemical phenotype and possible interactions with melanin-concentrating hormone neurons. Regul Pept 104:33–39

Beaulieu C, Somogyi P (1991) Enrichment of cholinergic synaptic terminals on GABAergic neurons and coexistence of immunoreactive GABA and choline acetyltransferase in the same synaptic terminals in the striate cortex of the cat. J Comp Neurol 304:666–680

Ben-Ari Y, Kanazawa I, Kelly JS (1976) Exclusively inhibitory action of iontophoretic acetylcholine on single neurones of feline thalamus. Nature 259:327–330

Berger H (1929) Über das electroenkephalogram des Menschen. Arch Psychiat Nervenkr 87:527–570

Berger B, Gaspar P (1994) Phylogeny and development of catecholamine systems in the CNS of vertebrates. Cambridge University Press, Cambridge (UK)

Berger B, Gaspar P, Verney C (1991) Dopaminergic innervation of the cerebral cortex: unexpected differences between rodents and primates. Trends Neurosci 14:21–27

Berlucchi G (1966) Electroencephalographic activity of the isolated hemicerebrum of the cat. Exp Neurol 15:220–228

Berlucchi G, Maffei L, Moruzzi G, Strata P (1964) EEG and behavioral effects elicited by cooling of medulla and pons. Arch Ital Biol 102:372–392

Berridge CW, Foote SL (1991) Effects of locus coeruleus activation on electroencephalographic activity in neocortex and hippocampus. J Neurosci 11:3135–3145

Bickford ME, Günlük AE, Guido W, Sherman SM (1993) Evidence that cholinergic axons from the parabrachial region of the brainstem are the exclusive source of nitric oxide in the lateral geniculate nucleus of the cat. J Comp Neurol 334:410–430

Bickford ME, Ramcharan E, Godwin DW, Erişir A, Gnadt J, Sherman SM (2000) Neurotransmitters contained in the subcortical extraretinal inputs to the monkey lateral geniculate nucleus. J Comp Neurol 424:701–717

Bonvallet M, Dell P (1965) Controle bulbaire du systéme activateur. In: Centre national de la recherche scientifique (France) Colloque international 127e, 1963 Lyon. Aspects anatomo-fonctionnels de la physiologie du someil. CNRS, Paris, pp 133–149

Bourgin P, Huitrón-Résendiz S, Spier AD, Fabre V, Morte B, Criado JR, Sutcliffe JG, Henriksen SJ, de Lecea L (2000) Hypocretin-1 modulates rapid eye movement sleep through activation of locus coeruleus neurons. J Neurosci 20:7760–7765

Bremer F (193a) Cerveau "isolé" et physiologie du sommeil. CR Soc Biol (Paris) 118:1235–1241

Bremer F (193b) L'activité cérébrale au cours du sommeil et de la narcose. Contribution à l'étude du mécanisme du sommeil. Bull Acad roy Méd Belg 4:68–86

Bremer F (193c) L'activité de l'écorce cérébrale et le problème physiologique du sommeil. Bull Soc Ital Biol Sper 13:271–290

Bremer F (1970) Preoptic hypnogenic focus and mesencephalic reticular formation. Brain Res 21:132–144

Bremer F (1951) Les réactions auditive de l'écorce cérébrale. L'année Psychol, 115–128

Brown RE, Sergeeva OA, Eriksson KS, Haas HL (200a) Convergent excitation of dorsal raphe serotonin neurons by multiple arousal systems (orexin/hypocretin, histamine, and noradrenaline). J Neurosci 22:8850–8859

Brown RE, McKenna JT, Winston S, Basheer R, Yanagawa Y, Thakkar MM, McCarley RW (200b) Characterization of GABAergic neurons in rapid-eye-movement sleep controlling regions of the brainstem reticular formation in GAD67-green fluorescent protein knock-in mice. Eur J Neurosci 27:352–363

Brown RE, Winston S, Basheer R, Thakkar MM, McCarley RW (2006) Electrophysiological characterization of neurons in the dorsolateral pontine rapid-eye-movement sleep induction zone of the rat: intrinsic membrane properties and responses to carbachol and orexins. Neuroscience 143:739–755

Buchwald NA, Wyers EJ, Okuma T, Heuser G (1961) The "caudate-spindle". I. Electrophysiological properties. Electroencephalogr Clin Neurophysiol 13:509–518

Burlet S, Tyler CJ, Leonard CS (2002) Direct and indirect excitation of laterodorsal tegmental neurons by Hypocretin/Orexin peptides: implications for wakefulness and narcolepsy. J Neurosci 22:2862–2872

Caballero A, De Andrés I (1986) Unilateral lesions in locus coeruleus area enhance paradoxical sleep. Electroencephalogr Clin Neurophysiol 64:339–346

Camacho-Evangelista A (1962) Modificaciones del electrocorticograma tras coagulaciones uni y bilaterales en el pedúnculo cerebeloso superior (un estudio experimental en el gato libre). An Anat 11:27–77

Camacho-Evangelista A, Reinoso-Surez F (1964) Activating and synchronizing centers in cat brain: electroencephalograms after lesions. Science 146:268–270

Camacho-Evangelista A, Reinoso-Surez F (1965) Effects on the EEG of lesions in the tractus brachium conjunctivum. 6th International congress of electroencephalography and clinical neurophysiology, pp 467–470

Chemelli RM, Willie JT, Sinton CM, Elmquist JK, Scammell T, Lee C, Richardson JA, Williams SC, Xiong Y, Kisanuki Y, Fitch TE, Nakazato M, Hammer RE, Saper CB, Yanagisawa M (1999) Narcolepsy in orexin knockout mice: molecular genetics of sleep regulation. Cell 98:437–451

Chou TC, Bjorkum AA, Gaus SE, Lu J, Scammell TE, Saper CB (2002) Afferents to the ventrolateral preoptic nucleus. J Neurosci 22:977–990

Ciliax BJ, Drash GW, Staley JK, Haber S, Mobley CJ, Miller GW, Mufson EJ, Mash DC, Levey AI (1999) Immunocytochemical localization of the dopamine transporter in human brain. J Comp Neurol 409:38–56

Cirelli C, Pompeiano M, Tononi G (1996) Neural gene expression in the waking state: a role for the locus coeruleus. Science 274:1211–1215

Clasc F, Llamas A, Reinoso-Surez F (1989) Hypothalamic connections of the insular cortex in the cat. Brain Res 490:361–366

Clements JR, Grant S (1990) Glutamate-like immunoreactivity in neurons of the laterodorsal tegmental and pedunculopontine nuclei in the rat. Neurosci Lett 120:70–73

Cordeau JP, Mancia M (1959) Evidence for the existence of an electroencephalographic synchronization mechanism originating in the lower brain stem. Electroencephalogr Clin Neurophysiol 11(3):551–564

Cordeau JP, Moreau A, Beaulnes A, Laurin C (1963) EEG and behavioral changes following microinjections of acetylcholine and adrenaline in the brain stem of cats. Arch Ital Biol 101:30–47

Coull JT, Frith CD, Dolan RHJ, Frackowiack RSJ, Grasby PM (1997) The neural correlates of the noradrenergic modulation of human attention, arousal and learning. Eur J Neurosci 9:589–598

Cox C, Sherman SM (1999) Glutamate inhibits thalamic reticular neurons. J Neurosci 19:6694–6699

Cunchillos JD, De Andrés I (1982) Participation of the cerebellum in the regulation of the sleep-wakefulness cycle. Results in cerebellectomized cats. Electroencephalogr Clin Neurophysiol 53:549–558

Curró-Dossi R, Paré D, Steriade M (1991) Short-lasting nicotinic and long-lasting muscarinic depolarizing responses of thalamocortical neurons to stimulation of mesopontine cholinergic nuclei. J Neurophysiol 65:393–406

Dahan L, Astier B, Vautrelle N, Urbain N, Kocsis B, Chouvet G (2007) Prominent burst firing of dopaminergic neurons in the ventral tegmental area during paradoxical sleep. Neuropsychopharmacology 32:1232–1241

Dahlström A, Fuxe K (1964) Evidence for the existence of monoamines-containing neurons in the central nervous system. I. Demonstration of monoamines in the cell bodies lf the brain stem neurons. Acta Physiol Scand [Suppl] 62:1–55

Davidson RA, Fedio P, Smith BD, Aureille E, Martin A (1992) Lateralized mediation of arousal and habituation: differential bilateral electrodermal activity in unilateral temporal lobectomy patients. Neuropsychologia 30:1053–1063

De Andrés I, Reinoso-Surez F (1979) Participation of the cerebellum in the regulation of the sleep-wakefulness cycle through the superior cerebellar peduncle. Arch Ital Biol 117:140–163

De Andrés I, Gutiérrez-Rivas E, Reinoso-Surez F (197a) Modifications of the sleep-wakefulness cycle after lesions at the level or oral pontine tegmentum and superior cerebellar peduncle. In: Levin P, Koella WP (eds) Sleep 1974. Karger Medical and Scientific Publishers, Basel, pp 235–238

De Andrés I, Gutiérrez-Rivas E, Nava E, Reinoso-Surez F (197b) Independence of sleep-wakefulness cycle in an implanted head 'encephale isole'. Neurosci Lett 2:13–18

De Andrés I, Gómez-Montoya J, Gutiérrez-Rivas E, Reinoso-Surez F (198a) Differential action upon sleep states of ventrolateral and central areas of pontine tegmental field. Arch Ital Biol 123:1–11

De Andrés I, Rodrigo-Angulo ML, Reinoso-Surez F (198b) El tegmento pontino: implicación en los estados de vigilia y de sueño. Rev Esp Fisiol Suppl 45:158–167

De la Roza C, Reinoso-Surez F (2006) GABAergic structures in the ventral part of the oral pontine reticular nucleus: an ultrastructural immunogold analysis. Neuroscience 142:1183–1193

De la Roza C, Martinez-Mena J, Sanchez-Valle ME, Reinoso-Suarez F (2004) Projections from the cat posterior lateral hypothalamus to the ventral part of the oral pontine reticular nucleus contain a GABAergic component. Brain Res 1020:118–129

de Lecea L, Sutcliffe JG (2005) The hypocretins and sleep. FEBS J 272:5675–5688

de Lecea L, Kilduff TS, Peyron C, Gao X, Foye PE, Danielson PE, Fukuhara C, Battenberg EL, Gautvik VT, Bartlett FS, Frankel WN, van den Pol AN, Bloom FE, Gautvik KM, Sutcliffe JG (1998) The hypocretins: hypothalamus-specific peptides with neuroexcitatory activity. Proc Natl Acad Sci USA 95:322–327

De Lima AD, Singer W (1987) The brainstem projection to the lateral geniculate nucleus in the cat: identification of cholinergic and monoaminergic elements. J Comp Neurol 259:92–121

De Lima AD, Montero VM, Singer W (1985) The cholinergic innervation of the visual thalamus: an EM immunocytochemical study. Exp Brain Res 59:206–212

Détri L, Semba K, Rasmusson DD (1997) Responses of cortical EEG-related basal forebrain neurons to brainstem and sensory stimulation in urethane-anaesthetized rats. Eur J Neurosci 9:1153–1161

Devilbiss DM, Waterhouse BD (2004) The effects of tonic locus ceruleus output on sensory-evoked responses of ventral posterior medial thalamic and barrel field cortical neurons in the awake rat. J Neurosci 24:10773–10785

Devilbiss DM, Page ME, Waterhouse BD (2006) Locus ceruleus regulates sensory encoding by neurons and networks in waking animals. J Neurosci 26:9860–9872

Divac I (1975) Magnocellular nuclei of the basal forebrain project to neocortex, brain stem, and olfactory bulb. Review of some functional correlates. Brain Res 93:385–398

Dringenberg HC, Kuo MC (2003) Histaminergic facilitation of electrocorticographic activation: role of basal forebrain, thalamus, and neocortex. Eur J Neurosci 18:2285–2291

Dzirasa K, Ribeiro S, Costa R, Santos LM, Lin SC, Grosmark A, Sotnikova TD, Gainetdinov RR, Caron MG, Nicolelis MA (2006) Dopaminergic control of sleep-wake states. J Neurosci 26:10577–10589

Edwards SB (1975) Autoradiographic studies of the projections of the midbrain reticular formation: descending projections of nucleus cuneiformis. J Comp Neurol 161:341–358

Eggermann E, Serafin M, Bayer L, Machard D, Saint-Mleux B, Jones BE, Mühlethaler M (2001) Orexins/hypocretins excite basal forebrain cholinergic neurons. Neuroscience 108:177–181

Eisensehr I, Noachtar S, von Schlippenbach C, Uttner I, Kleine J, Seelos K, Helmchen C (2003) Hypersomnia associated with bilateral posterior hypothalamic lesion. A polysomnographic case study. Eur Neurol 49:169–172

el Mansari M, Sakai K, Jouvet M (1989) Unitary characteristics of presumptive cholinergic tegmental neurons during the sleep-waking cycle in freely moving cats. Exp Brain Res 76:519–529

España RA, Baldo BA, Kelley AE, Berridge CW (2001) Wake-promoting and sleep-suppressing actions of hypocretin (orexin): basal forebrain sites of action. Neuroscience 106:699–715

Fairén A (1973) Conexiones ascendentes del tegmento pontino caudal: Un estudio experimental en la rata blanca. Anal Anat 22:463–508

Findlay AL, Hayward JN (1969) Spontaneous activity of single neurones in the hypothalamus of rabbits during sleep and waking. J Physiol 201:237–258

Fink RP, Heimer L (1967) Two methods for selective silver impregnation of degenerating axons and their synaptic endings in the central nervous system. Brain Res 4:369–374

Fisher RS, Buchwald NA, Hull CD, Levine MS (1988) GABAergic basal forebrain neurons project to the neocortex: the localization of glutamic acid decarboxylase and choline acetyltransferase in feline corticopetal neurons. J Comp Neurol 272:489–502

Fitzpatrick D, Diamond IT, Raczkowski D (1989) Cholinergic and monoaminergic innervation of the cat's thalamus: comparison of the lateral geniculate nucleus with other principal sensory nuclei. J Comp Neurol 288:647–675

Foote SL, Morrison JH (1987) Extrathalamic modulation of cortical function. Ann Rev Neurosci 10:41–66

Forcadas MI, Zarranz JJ (1994) Insomnia and hallucinations caused by vascular lesions of the pontine tegmentum. Neurologia 9:211–223

Fournier GN, Materi LM, Semba K, Rasmusson DD (2004) Cortical acetylcholine release and electroencephalogram activation evoked by ionotropic glutamate receptor agonists in the rat basal forebrain. Neuroscience 123:785–792

French JD, Magoun HW (1952) Effects of chronic lesions in central cephalic brain stem of monkeys. AMA Arch Neurol Psychiatry 68:591–604

Gallopin T, Fort P, Eggermann E, Cauli B, Luppi PH, Rossier J, Audinat E, Mühlethaler M, Serafin M (2000) Identification of sleep-promoting neurons in vitro. Nature 404:992–995

García Uría J, De Andrés I, Cunchillos JD, Reinoso-Surez F (1980) Modification of the sleep-wakefulness cycle after surgical removal of the cerebellar cortex. In: Popovicin L, Asgian B, Badin G (eds) Sleep l978. Karger Medical and Scientific Publishers, Basel, pp 289–292

García-Cabezas MA, Rico B, Snchez-Gonzlez MA, Cavada C (200a) Distribution of the dopamine innervation in the macaque and human thalamus. Neuroimage 34:965–984

García-Cabezas MA, Martinez-Sanchez P, Snchez-Gonzlez MA, Garzón M, Cavada C (200b) Dopamine innervation in the thalamus: monkey versus rat. Cereb Cortex 19:424–434

Garduño-Torres B, Treviño M, Gutiérrez R, Arias-Montaño JA (2007) Pre-synaptic histamine H3 receptors regulate glutamate, but not GABA release in rat thalamus. Neuropharmacology 52:527–535

Garzón M (1996) Estudio morfofuncional de los núcleos reticular oral y reticular caudal del tegmento pontino como regiones generadoras de sueño paradójico. Tesis Doctoral, Universidad Autónoma de Madrid
Gastaut H, Bert J (1961) Electroencephalographic detection of sleep induced by repetitive sensory stimuli. In: Wolstenholme GEW, O'Connor CM (eds) On the nature of the sleep. Churchil, London, pp 260–283
Gervasoni D, Darracq L, Fort P, Soulière F, Chouvet G, Luppi PH (1998) Electrophysiological evidence that noradrenergic neurons of the rat locus coeruleus are tonically inhibited by GABA during sleep. Eur J Neurosci 10:964–970
Gervasoni D, Peyron Ch, Rampon C, Barbagli B, Chouvet G, Urbain N, Fort P, Luppi PH (2000) Role and origin of the GABAergic innervation of dorsal raphe. Serotonergic neurons. J Neurosci 20:4217–4225
Goldman-Rakic PS, Brown RM (1982) Postnatal development of monoamine content and synthesis in the cerebral cortex of rhesus monkeys. Brain Res 256:339–349
Gottesmann C (1999) The neurophysiology of sleep and waking: intracerebral connections, functioning and ascending influences of the medulla oblongata. Prog Neurobiol 59:1–54
Gottesmann C, Gandolfo G, Zernicki B (1995) Sleep-waking cycle in chronic rat preparations with brain stem transected at the caudopontine level. Brain Res Bull 36:573–580
Govindaiah G, Cox CL (2006) Modulation of thalamic neuron excitability by orexins. Neuropharmacology 51:414–425
Graybiel AM (1977) Direct and indirect preoculomotor pathways of the brainstem: an autoradiographic study of the pontine reticular formation in the cat. J Comp Neurol 175:37–78
Graybiel AM, Berson DM (1980) Histochemical identification and afferent connections of subdivisions in the lateralis posterior-pulvinar complex and related thalamic nuclei in the cat. Neuroscience 5:1175–1238
Gritti I, Mainville L, Mancia M, Jones BE (1997) GABAergic and other non-cholinergic basal forebrain neurons project together with cholinergic neurons to meso- and iso-cortex in the rat. J Comp Neurol 383:163–177
Grofov I, Ottersen OP, Rinvik E (1978) Mesencephalic and diencephalic afferents to the superior colliculus and periaqueductal gray substance demonstrated by retrograde axonal transport of horseradish peroxidase in the cat. Brain Res 146:205–220
Gutiérrez-Rivas E, De Andrés I, Gómez-Montoya J, Reinoso-Surez F (1978) The influence of the rostropontine-ventrolateral region on the sleep-wakefulness cycle. Experientia 34:61–62
Haas H, Panula P (2003) The role of histamine and the tuberomamillary nucleus in the nervous system. Nat Rev 4:121–130
Hallanger AE, Levey AI, Lee HJ, Rye DB, Wainer BH (1987) The origins of cholinergic and other subcortical afferents to the thalamus in the rat. J Comp Neurol 262:105–124
Harting JK, Van Lieshout DP, Hashikawa T, Weber JT (1991) The parabigeminogeniculate projection: connectional studies in eight mammals. J Comp Neurol 305:559–581
Heckers S, Geula C, Mesulam MM (1992) Cholinergic innervation of the human thalamus: dual origin and differential nuclear distribution. J Comp Neurol 325:68–82
Henny P, Jones BE (2008) Projections from basal forebrain to prefrontal cortex comprise cholinergic, GABAergic and glutamatergic inputs to pyramidal cells or interneurons. Eur J Neurosci 27:654–670
Hernndez-Peón R (1962) Sleep induced by localized electrical or chemical stimulation of the forebrain. Electroencephalogr Clin Neurophysiol 14:423–424
Hess WR (1927) Stammganglien Reizvesuche. Ber ges Physiol 42:554–555
Hess WR (1931) Le sommeil. C R Soc Biol, Paris 107:1333–1360
Hess WR (1968) Psychologie in biologisher Sicht. 2 Aufl Stuttgart, G Thieme
Hodes R (1964) Electrocortical desynchronization resulting from spinal block: evidence for synchronizing influences in the cervical cord. Arch Ital Biol 102:183–196

Holmes CJ, Jones BE (1994) Importance of cholinergic, GABAergic, serotonergic and other neurons in the medial medullary reticular formation for sleep-wake states studied by cytotoxic lesions in the cat. Neuroscience 62:1179–1200

Holstege G, Kuypers HG (1982) The anatomy of brain stem pathways to the spinal cord in cat. A labeled amino acid tracing study. Prog Brain Res 57:145–175

Horvath FE, Buser P (1972) Thalamo-caudate-cortical relationships in synchronized activity. I. Differentiation between ventral and dorsal spindle systems. Brain Res 39:21–41

Huang ZL, Qu WM, Li WD, Mochizuki T, Eguchi N, Watanabe T, Urade Y, Hayaishi O (200a) Arousal effect of orexin A depends on activation of the histaminergic system. Proc Natl Acad Sci USA 98:9965–9970

Huang H, Ghosh P, van den Pol AN (200b) Prefrontal cortex-projecting glutamatergic thalamic paraventricular nucleus-excited by hypocretin: a feedforward circuit that may enhance cognitive arousal. J Neurophysiol 95:1656–1668

Inagaki N, Yamatodani A, Ando-Yamamoto M, Tohyama M, Watanabe T, Wada H (1988) Organization of histaminergic fibers in rat brain. J Comp Neurol 273:283–300

Ingram WR, Barris RW, Ranson SW (1936) Catalepsy. An experimental study. Arch Neurol Psychiat 35:1175–1197

Jacobs BL, Wilkinson LO, Fornal CA (1990) The role of brain serotonin. A neurophysiologic perspective. Neuropsychopharmacology 3:473–479

Jiménez-Castellanos J, Reinoso-Surez F (1985) Topographical organization of the afferent connections of the principal ventro-medial thalamic nucleus in the cat. J Comp Neurol 236:297–314

Jin CY, Panula P (2005) The laminar histamine receptor system in human prefrontal cortex suggests multiple levels of histaminergic regulation. Neuroscience 132:137–149

Jin CY, Kalimo H, Panula P (2002) The histaminergic system in human thalamus: correlation of innervation to receptor expression. Eur J Neurosci 15:1125–1138

Jones EG (1975) Some aspects of the organization of the thalamic reticular complex. J Comp Neurol 162:285–308

Jones BE (1991) Paradoxical sleep and its chemical/structural substrates in the brain. Neuroscience 40:637–656

Jones EG (2001) The thalamic matrix and thalamocortical synchrony. Trends Neurosci 24:595–601

Jones EG (2007) The thalamus. Cambridge University Press, Cambridge, UK

Jones BE (2008) Modulation of cortical activation and behavioral arousal by cholinergic and orexinergic systems. Ann NY Acad Sci 1129:26–34

Jones BE, Beaudet A (1987) Retrograde labeling of neurones in the brain stem following injections of [3H]choline into the forebrain of the rat. Exp Brain Res 65:437–448

Jones B, Bobillier P, Jouvet M (1969) Effects de la destruction des neurones contenant des catecolamines du mesencephale sur le cycle veille-sommeil du chat. C R Soc Biol (Paris) 163:176–180

Jouvet M (1962) Research on the neural structures and responsible mechanisms in different phases of physiological sleep. Arch Ital Biol 100:125–206

Karczmar AG, Longo VG, Scotti deCarolis A (1970) A pharmacological model of paradoxical sleep: the role of cholinergic and monoamine systems. Physiol Behav 5:175–182

Kayama Y, Koyama Y (2003) Control of sleep and wakefulness by brainstem monoaminergic and cholinergic neurons. Acta Neurochir Suppl 87:3–6

Kayama Y, Ohta M, Jodo E (1992) Firing of 'possibly' cholinergic neurons in the rat laterodorsal tegmental nucleus during sleep and wakefulness. Brain Res 569:210–220

Keifer JC, Baghdoyan HA, Lydic R (1996) Pontine cholinergic mechanisms modulate the cortical electroencephalographic spindles of halothane anesthesia. Anesthesiology 84:945–954

Khateb A, Fort P, Pegna A, Jones BE, Miihlethaler M (1995) Cholinergic nucleus basalis neurons are excited by histamine in vitro. Neuroscience 69:495–506

Kievit J, Kuypers HG (1975a) Basal forebrain and hypothalamic connection to frontal and parietal cortex in the Rhesus monkey. Science 187:660–662
Kievit J, Kuypers HG (1975b) Subcortical afferents to the frontal lobe in the rhesus monkey studied by means of retrograde horseradish peroxidase transport. Brain Res 85:261–266
Kinomura S, Larsson J, Gulyas B, Roland PE (1996) Activation by attention of the human reticular formation and thalamic ntralaminar nuclei. Science 271:512–515
Koyama Y, Kayama Y (1993) Mutual interactions among cholinergic, noradrenergic and serotonergic neurons studied by ionophoresis of these transmitters in rat brainstem nuclei. Neuroscience 55:1117–1126
Koyama Y, Takahashi K, Kodama T, Kayama Y (2003) State-dependent activity of neurons in the perifornical hypothalamic area during sleep and waking. Neuroscience 119:1209–1219
Kruglikov I, Rudy B (2008) Perisomatic GABA release and thalamocortical integration onto neocortical excitatory cells are regulated by neuromodulators. Neuron 58:911–924
Kubota Y, Hatada S, Kondo S, Karube F, Kawaguchi Y (2007) Neocortical inhibitory terminals innervate dendritic spines targeted by thalamocortical afferents. J Neurosci 27:1139–1150
Kumar S, Szymusiak R, Bashir T, Suntsova N, Rai S, McGinty D, Alam MN (2008) Inactivation of median preoptic nucleus causes c-Fos expression in hypocretin- and serotonin-containing neurons in anesthetized rat. Brain Res 1234:66–77
Kuo MC, Dringenberg HC (2008) Histamine facilitates in vivo thalamocortical long-term potentiation in the mature visual cortex of anesthetized rats. Eur J Neurosci 27:1731–1738
Lambe E, Aghajanian G (2003) Hypocretin (orexin) induces calcium transients in single spines postsynaptic to identified thalamocortical boutons in prefrontal slice. Neuron 40:139–150
Lambe EK, Olausson P, Horst NK, Taylor JR, Aghajanian GK (2005) Hypocretin and nicotine excite the same thalamocortical synapses in prefrontal cortex: correlation with improved attention in rat. J Neurosci 25:5225–5229
Lambe EK, Liu RJ, Aghajanian GK (2007) Schizophrenia, hypocretin (orexin), and the thalamocortical activating system. Schizophr Bull 33:1284–1290
Larsen JP, Tandberg E (2001) Sleep disorders in patients with Parkinson's disease: epidemiology and management. CNS Drugs 15:267–275
Lavoie B, Parent A (1994) Pedunculopontine nucleus in the squirrel monkey: distribution of cholinergic and monoaminergic neurons in the mesopontine tegmentum with evidence for the presence of glutamate in cholinergic neurons. J Comp Neurol 344:190–209
Lee KH, McCormick DA (1995) Acetylcholine excites GABAergic neurons of the ferret perigeniculate nucleus through nicotinic receptors. J Neurophysiol 73:2123–2128
Lee RS, Steffensen SC, Henriksen SJ (2001) Discharge profiles of ventral tegmental area GABA neurons during movement, anesthesia, and the sleep-wake cycle. J Neurosci 21:1757–1766
Lee KH, Broberger C, Kim U, McCormick DA (2004) Histamine modulates thalamocortical activity by activating a chloride conductance in ferret perigeniculate neurons. Proc Natl Acad Sci USA 101:6716–6721
Lee MG, Hassani OK, Jones BE (2005a) Discharge of identified orexin/hypocretin neurons across the sleep-waking cycle. J Neurosci 25:6716–6720
Lee MG, Hassani OK, Alonso A, Jones BE (2005b) Cholinergic basal forebrain neurons burst with theta during waking and paradoxical sleep. J Neurosci 25:4365–4369
Leonard CS, Llins R (1994) Serotonergic and cholinergic inhibition of mesopontine cholinergic neurons controlling REM sleep: an in vitro electrophysiological study. Neuroscience 59:309–330
Levey AI, Hallanger AE, Wainer BH (1987) Choline acetyltransferase immunoreactivity in the rat thalamus. J Comp Neurol 257:317–332
Lewis DA, Melchitzky DS, Sesack SR, Whitehead RE, Auh S, Sampson A (2001) Dopamine transporter immunoreactivity in monkey cerebral cortex: regional, laminar, and ultrastructural localization. J Comp Neurol 432:119–136
Lin JS (1994) The brain histaminergic system and arousal mechanisms. Snews Alert:11

Lin JS (2000) Brain structures and mechanisms involved in the control of cortical activation and wakefulness, with emphasis on the posterior hypothalamus and histaminergic neurons. Sleep Med Rev 4:471–503
Lin JS, Sakai K, Jouvet M (1986) Role of hypothalamic histaminergic systems in the regulation of the states of vigilance in the cat. CR Acad Sci 303:469–474
Lin JS, Sakai K, Jouvet M (1988) Evidence for histaminergic arousal mechanisms in the hypothalamus of cats. Neuropharmacology 27:111–122
Lin JS, Sakai K, Vanni-Mercier G, Jouvet M (1989) A critical role of the posterior hypothalamus in the mechanisms of wakefulness determined by microinjection of muscimol in freely moving cats. Brain Res 479(2):225–240
Lin JS, Kitahama K, Fort P, Panula P, Denney RM, Jouvet M (1993) Histaminergic system in the posterior hypothalamus in the cat with reference to type B monoamine oxidase. J Comp Neurol 330:405–420
Lin JS, Hou Y, Sakai K, Jouvet M (1996) Histaminergic descending inputs to the mesopontine tegmentum and their role in the control of cortical activation and wakefulness in the cat. J Neurosci 16:1523–1537
Lindsley DB, Schreiner LH, Knowles WB, Magoun HW (1950) Behavioral and EEG changes following chronic brain stem lesions in the cat. Electroencephalogr Clin Neurophysiol 2:483–498
Llamas A (1966) Conexiones de la parte compacta de la sustancia negra y del rea ventral tegmrental de Tsai. An Anat 18:355–392
Llamas A, Reinoso-Surez F (1969) Projections of the substantia nigra and ventral tegmental mesencephalic area. In: Gilligham FG (ed) Third symposium on Parkinson's disease. E. & S. Livingstone, London, pp 82–97
Llamas A, Reinoso-Surez F, Martínez-Moreno E (1975) Projections to the gyrus proreus from the brain stem tegmentum (locus coeruleus, raphe nuclei) in the cat, demonstrated by retrorade transport of horseradish peroxidase. Brain Res 89:331–336
Losier BJ, Semba K (1993) Dual projections of single cholinergic and aminergic brainstem neurons to the thalamus and basal forebrain in the rat. Brain Res 604:41–52
Lu J, Bjorkum AA, Xu M, Gaus SE, Shiromani PJ, Saper CB (2002) Selective activation of the extended ventrolateral preoptic nucleus during rapid eye movement sleep. J Neurosci 22:4568–4576
Lu J, Jhou TC, Saper CB (2006) Identification of wake-active dopaminergic neurons in the ventral periaqueductal gray matter. J Neurosci 26:193–202
Luebke JI, Greene RW, Semba K, Kamondi A, McCarley RW, Reiner PB (1992) Serotonin hyperpolarizes cholinergic low-threshold burst neurons in the rat laterodorsal tegmental nucleus in vitro. Proc Natl Acad Sci USA 89:743–747
Lugaresi E, Medori R, Montagna P, Baruzzi A, Cortelli P, Lugaresi A, Tinuper P, Zucconi M, Gambetti P (1986) Fatal familial insomnia and dysautonomia with selective degeneration of thalamic nuclei. N Engl J Med 315:997–1003
Luppi P-H, Peyron C, Rampon C, Gervasont D, Barbagli B, Boisard B, Fort P (1998) Inhibitory mechanisms in dorsal raphe nucleus and locus coeruleus during sleep. In: Lydic R, Baghdoyan HA (eds) Handbook of behavioural state control. Cellular and molecular mechanisms. CRC Press, Boca Raton, London, New York, pp 195–211
Madoz P (1968) Influencia de la región preóptica en la regulación de la actividad eléctrica cerebral. Tesis Doctoral, Universidad de Navarra, Pamplona, Spain
Madoz P, Reinoso-Surez F (1968) Influence of lesions in preoptic region on the states of sleep and wakefulness. Proc XXIV Int Cong Physio Sci 7:276
Magnes J, Moruzzi G, Pompeiano O (1961) Synchronization of the EEG produced by low frequency electrical stimulation of the region of the solitary tract. Arch Ital Biol 99:33–77
Magni F, Moruzzi G, Rossi GF, Zanchetti A (1959) EEG arousal following inactivation of the lower brain stem by selective injection of barbiturate into the vertebral circulation. Arch Ital Biol 97:33–46

Maloney KJ, Mainville L, Jones B (1999) Differential c-Fos expression in cholinergic, monoaminergic, and GABAergic cell groups of the pontomesencephalic tegmentum after paradoxical sleep deprivation and recovery. J Neurosci 15:3057–3072

Mancia M (1969) Electrophysiological and behavioural changes owing to splitting of the brainstem in cats. Electroencephalogr Clin Neurophysiol 27:487–502

Mancia M, Meulders M, Santibanez HG (1959) Synchronisation de l'électroencéphalogramme provoquée par la stimulation visuelle répétitive chez le chat "médiopontin prétrigéminal". Arch Int Physiol Biochem 67:661–670

Mancia M, Margnelli M, Mariotti M, Spreafico R, Broggi G (1974a) Brain stem-thalamus reciprocal influences in the cat. Brain Res 69:297–314

Mancia M, Mariotti M, Spreafico R (1974b) Caudo-rostral brain stem reciprocal influences in the cat. Brain Res 80:41–51

Manns ID, Alonso A, Jones BE (2000) Discharge profiles of juxtacellularly labeled and immunohistochemically identified GABAergic basal forebrain neurons recorded in association with the electroencephalogram in anesthetized rats. J Neurosci 20:9252–9263

Marcus JN, Aschkenasi CJ, Lee CE, Chemelli RM, Saper CB, Yanagisawa M, Elmquist JK (2001) Differential expression of orexin receptors 1 and 2 in the rat brain. J Comp Neurol 435:6–25

Marini G, Gritti I, Mancia M (1988) Ibotenic acid lesions of the thalamic medialis dorsalis nucleus in the cat: effects on the sleep-waking cycle. Neurosci Lett 89:259–264

Marrocco RT, Witte EA, Davidson MC (1994) Arousal systems. Curr Opin Neurobiol 4:166–170

Masri R, Trageser JC, Bezdudnaya T, Li Y, Keller A (2006) Cholinergic regulation of the posterior medial thalamic nucleus. J Neurophysiol 96:2265–2273

McCormick DA (1992) Neurotransmitter actions in the thalamus and cerebral cortex and their role in neuromodulation of thalamocortical activity. Prog Neurobiol 39:337–388

McCormick DA, Bal T (1994) Sensory gating mechanisms of the thalamus. Curr Opin Neurobiol 4:550–556

McCormick DA, Bal T (1997) Sleep and arousal: thalamocortical mechanisms. Annu Rev Neurosci 20:185–215

McCormick DA, Pape HC (1988) Acetylcholine inhibits identified interneurons in the cat lateral geniculate nucleus. Nature 334:246–248

McCormick DA, Williamson A (1991) Modulation of neuronal firing mode in cat and guinea pig LGNd by histamine: possible cellular mechanisms of histaminergic control of arousal. J Neurosci 11:3188–3199

McCormick DA, Wang Z, Huguenard J (1993) Neurotransmitter control of neocortical neuronal activity and excitability. Cereb Cortex 3:387–398

McGinty DJ, Harper RM (1976) Dorsal raphe neurons: depression of firing during sleep in cats. Brain Res 101:569–575

McGinty DJ, Sterman MB (1968) Sleep suppression after basal forebrain lesions in the cat. Science 160:1253–1255

Mesulam MM (1995) Cholinergic pathways and the ascending reticular activating system of the human brain. Ann NY Acad Sci 757:169–179

Mesulam MM, Geula C (1988) Nucleus basalis (Ch4) and cortical cholinergic innervation in the human brain: observations based on the distribution of acetylcholinesterase and choline acetyltransferase. J Comp Neurol 275:216–240

Mileykovskiy BY, Kiyashchenko LI, Siegel JM (2005) Behavioral correlates of activity in identified hypocretin/orexin neurons. Neuron 46:787–798

Miller JD, Farber J, Gatz P, Roffwarg H, German DC (1983) Activity of mesencephalic dopamine and non-dopamine neurons across stages of sleep and walking in the rat. Brain Res 273:133–141

Monti JM, Monti D (2007) The involvement of dopamine in the modulation of sleep and waking. Sleep Med Rev 11:113–133

Monti JM, Jantos H, Boussard M, Altier H, Orellana C, Olivera S (1991) Effects of selective activation or blockade of the histamine H, receptor on sleep and wakefulness. Eur J Pharmacol 205:283–287

Moore RY, Bloom FE (1979) Central catecholamine neuron systems: anatomy and physiology of the norepinephrine and epinephrine systems. Annu Rev Neurosci 2:113–168

Moore RY, Abrahamson EA, Van Den Pol A (2001) The hypocretin neuron system: an arousal system in the human brain. Arch Ital Biol 139:195–205

Moreno-Balandrn E, Garzón M, Bódalo C, Reinoso-Surez F, De Andrés I (2008) Sleep-wakefulness effects after microinjections of hypocretin 1 (orexin A) in cholinoceptive areas of the cat oral pontine tegmentum. Eur J Neurosci 28:331–341

Morrison RS, Dempsey EW (1942) A study of thalamo-cortical relations. Am J Physiol 135:282–302

Morrison JH, Foote SL (1986) Noradrenergic and serotoninergic innervation of cortical, thalamic, and tectal visual structures in Old and New World monkeys. J Comp Neurol 243:117–138

Moruzzi G (1972) The sleep-waking cycle. Ergeb Physiol 64:1–165

Moruzzi G, Magoun HW (1949) Brain stem reticular formation and activation of the EEG. Electroenceph Clin Neurophysiol 1:455–473

Murillo-Rodriguez E, Liu M, Blanco-Centurion C, Shiromani PJ (2008) Effects of hypocretin (orexin) neuronal loss on sleep and extracellular adenosine levels in the rat basal forebrain. Eur J Neurosci 28:1191–1198

Nauta WJH (1946) Hypothalamic regulation of sleep in rats. An experimental study. J Neurophysiol 9:285–316

Nauta WJH (1957) Silver impregnation of degenerating axons. In: Windle WF, Thomas F (eds) New research techniques in neuroanatomy. Charles C. Thomas, Springfield, pp 17–26

Nauta WJH, Haymaker W (1969) Hypothalamic nuclei and fiber connections. In: Haymaker W, Anderson E, Nauta WJH (eds) The hypothalamus. Charles C Thomas, Springfield, pp 136–209

Nauta WJH, Kuypers HGJ (1958) Some ascending pathways in the brain stem reticular formation. In: Reticular formation of the brain. Henry Ford Hospital symposium. Little Brown, Boston, pp 3–30

Nelson JP, McCarley RW, Hobson JA (1983) REM sleep burst neurons, PGO waves, and eye movement information. J Neurophysiol 50:784–797

Nicholson AN (1983) Antihistamines and sedation. Lancet 2:211–212

Nicholson C (2007) Light sleeper. Nat Rev Neurosci 8:911

Nicolelis MA, Fanselow EE (2002) Thalamocortical [correction of Thalamcortical] optimization of tactile processing according to behavioral state. Nat Neurosci 5:517–523. Erratum in: Nat Neurosci 5:704

Nicolelis MA, Chapin JK, Lin RC (1992) Somatotopic maps within the zona incerta relay parallel GABAergic somatosensory pathways to the neocortex, superior colliculus, and brainstem. Brain Res 577:134–141

Nitz D, Siegel JM (1997) GABA release in the locus coeruleus as a function of sleep/wake state. Neuroscience 78:795–801

Nuñez A, Buño W, Reinoso-Surez F (1998) Neurotransmitter actions on oral pontine tegmental neurons of the rat: an in vitro study. Brain Res 804:144–148

Nuñez A, Moreno-Balandrn ME, Rodrigo-Angulo ML, Garzón M, de Andrés I (2006) Relationship between the perifornical hypothalamic area and the oral pontine reticular nucleus in the rat. Possible implication of the hypocretinergic projection in the control of rapid eye movement sleep. Eur J Neurosci 24:2834–2842

Nuñez A, Rodrigo-Angulo ML, Andrés ID, Garzón M (2009) Hypocretin/Orexin neuropeptides: participation in the control of sleep-wakefulness cycle and energy homeostasis. Curr Neuropharmacol 7:50–59

Ojima H (1994) Terminal morphology and distribution of corticothalamic fibers originating from layers 5 and 6 of cat primary auditory cortex. Cereb Cortex 4:646–663
Oka H, Ito J, Kawamura M (1982) Identification of thalamo-cortical neurons responsible for cortical recruiting and spindling activities in cats. Neurosci Lett 33:13–18
Onoé H, Yamatodani A, Watanabe Y, Mochizuki T, Wada H, Hayashi O (1992) Prostaglandin E2 and histamine in the posterior hypothalamus. J Sleep Res l(Suppl 1):166
Otake K, Ruggiero DA (1995) Monoamines and nitric oxide are employed by afferents engaged in midline thalamic regulation. J Neurosci 15:1891–1911
Panula P, Pirvola U, Auvinen S, Airaksinen MS (1989) Histamineimmunoreactive nerve fibers in the rat brain. Neuroscience 28:585–610
Pape HC, McCormick DA (1995) Electrophysiological and pharmacological properties of interneurons in the cat dorsal lateral geniculate nucleus. Neuroscience 68:1105–1125
Parent A, Steriade M (1984) Midbrain tegmental projections of nucleus reticularis thalami of cat and monkey: a retrograde transport and antidromic invasion study. J Comp Neurol 229:548–558
Parmentier R, Ohtsu H, Djebbara-Hannas Z, Valatx JL, Watanabe T, Lin JS (2002) Anatomical, physiological, and pharmacological characteristics of histidine decarboxylase knock-out mice: evidence for the role of brain histamine in behavioral and sleep-wake control. J Neurosci 22:7695–7711
Pasquier DA, Reinoso-Surez F (1976) Direct projections from hypothalamus to hippocampus in the rat demonstrated by retrograde transport of horseradish peroxidase. Brain Res 108:165–169
Pasquier DA, Reinoso-Surez F (1977) Differential efferent connections of the brain stem to the hippocampus in the cat. Brain Res 120:540–548
Pasquier DA, Reinoso-Surez F (1978) The topographic organization of hypothalamic and brain stem projections to the hippocampus. Brain Res Bull 3:373–389
Patel NC, Carden WB, Bickford ME (1999) Synaptic targets of cholinergic terminals in the cat lateral posterior nucleus. J Comp Neurol 410:31–41
Paxinos G, Watson C (1986) The rat brain in steretaxic coordinates. Academic, San Diego
Peyron C, Tighe DK, van den Pol AN, de Lecea L, Heller HC, Sutcliffe JG, Kilduff TS (1998) Neurons containing hypocretin (orexin) project to multiple neuronal systems. J Neurosci 18:9996–10015
Peyron C, Faraco J, Rogers W, Ripley B, Overeem S, Charnay Y, Nevsimalova S, Aldrich M, Reynolds D, Albin R, Li R, Hungs M, Pedrazzoli M, Padigaru M, Kucherlapati M, Fan J, Maki R, Lammers GJ, Bouras C, Kucherlapati R, Nishino S, Mignot E (2000) A mutation in a case of early onset narcolepsy and a generalized absence of hypocretin peptides in human narcoleptic brains. Nat Med 9:991–997
Pompeiano O, Swett JE (1962) EEG behavioural manifestation of sleep induced by cutaneous nerve stimulation in normal cats. Arch Ital Biol 100:311–342
Puizillout JJ, Foutz AS, Ternaux JP (1973a) Slow wave and paradoxical sleep in 'encéphale isolé' preparation. In: Levin P, Koella WP (eds) Sleep 1973. Karger Medical and Scientific Publishers, Basel, pp 322–335
Puizillout JJ, Ternaux JP, Foutz AS, Dell P (1973b) Slow wave sleep with phasic discharges. Triggering by vago-aortic stimulation. Rev Electroencephalogr Neurophysiol Clin 3:21–37
Puizillout JJ, Ternaux JP, Foutz AS, Fernandez G (1974) Sleep stages in "encéphale isolé" preparations: I. Triggering of ponto-geniculo-occipital spikes and slow-wave sleep. The role of the nuclei of the raphe. Electroencephalogr Clin Neurophysiol 37:561–576
Ranson SW (1939) Somnolence caused by hypothalamic lesions in the monkey. Arch Neurol Psychiat 41:1–23
Rasmusson DD, Szerb IC, Jordan JL (1996) Differential effects of alpha-amino-3-hydroxy-5-methyl-4-isoxazole propionic acid and N-methyl-D-aspartate receptor antagonists applied to

the basal forebrain on cortical acetylcholine release and electroencephalogram desynchronization. Neuroscience 72:419–427

Rausell E, Avendaño C (1985) Thalamocortical neurons projecting to superficial and to deep layers in parietal, frontal and prefrontal regions in the cat. Brain Res 347:159–165

Rechtschaffen A, Kales A (1968) A manual of standardized terminology, techniques and scoring system for sleep stages of human subjects. University of California, Brain Information Service/Brain Research Institute, Los Angeles, CA

Reiner PB, Kamondi A (1994) Mechanisms of antihistamine-induced sedation in the human brain: H, receptor activation reduces a background leakage potassium current. Neuroscience 59:579–588

Reinoso-Barbero F, De Andrés I (1995) Effects of opioid microinjections in the nucleus of the solitary tract on the sleep-wakefulness cycle states in the cats. Anestesiology 82:144–152

Reinoso-Surez F (1952) El problema facilitación supresión en sistema nervioso central bajo un punto de vista anatómico. An Anat 1:59–84

Reinoso-Surez F (1954) Die Auswirkungen der Ausschaltung eines Nucleus ruber auf die Hirnrinde (Electroencephalographische Untersuchungen an der Katze). Dtsch Z Nervenheilk 172:201–219

Reinoso-Surez F (1959) Formación reticular y actividad bioeléctrica encefálica. I Modificación de la actividad bioeléctrica espontnea tras diatermocoagulaciones localizadas en diencéfalo y corteza frontal. An Anat 7:239–265

Reinoso-Surez F (1960) Modifications of the potential evoked by acoustic stimuli by means of diatermo-coagulation in the diencephalon. J Comp Neurol 114:207–215

Reinoso-Surez F (1961) Topographischer Hirnatlas der Katze fur Experimental-Physiologische Untersuchungen. Merck AG, Darmstad

Reinoso-Surez F (1962) Consideraciones sobre las bases anatomofisiológicas del control de las sensaciones a la luz de las modificaciones que sufren los potenciales evocados en las vías visual y auditiva tras lesiones subcorticales. Rev Med Univ Navarra 6:203–216

Reinoso-Surez F (1963) Effects of diencephalic and mesencepalic lesions upon afferent impulses in the central sensory pathways. Electroenceph Clin Neurophysiol Suppl 24:33–42

Reinoso-Surez F (1971) Centros reguladores de la vigilia y el sueño. Rev Esp Oto-Neuro-Oftalm 29:111–119

Reinoso-Surez F (1977) Proyecciones a neocortex e hipocampo desde el tronco del encéfalo e hipotlamo. An Anat Nº Extraord:1–9

Reinoso-Surez F (1984) Connectional patterns in parieto-temporo-occipital association cortex of the feline cerebral cortex. In: Reinoso-Surez F, Ajmone Marsan C (eds) Cortical integration: basic, archicortical and cortical association levels of neural integration, vol 11, IBRO monograph series. Raven Press, New York, pp 255–278

Reinoso-Surez F (1985) Morfología del sistema neuroendocrino. In: Schiaffini O, Fernndez-Tresguerrres JA (eds) Neuroendocrinología. Salvat Editores, Barcelona, pp 1–46

Reinoso-Surez F (1992) Considerations on thalamic and pontine sleep mechanisms. In: Velluti R (ed) Fundamental neurobiology. Work, Montevideo, pp 59–71

Reinoso-Surez F (1993) Algunas consideraciones sobre la anatomía del sueño. An Anat 39:165–181

Reinoso-Surez F (1997) Neurobiología del despertar y la vigilia. An R Acad Nac Med (Madr) 114:249–265

Reinoso-Surez F (2005) Neurobiología del sueño. Rev Med Univ Navarra 49:8–17

Reinoso-Surez F, De Andrés I (1976) Brain structures and sleep. Trab Inst Cajal Invest Biol 68:39–68

Reinoso-Surez F, Llamas A (1968) Fibras ascendentes desde el tegmento pontino oral en el ratón. Acta Neurol Latinoam 14:5–16

Reinoso-Surez F, Llamas A (1975) Conexiones aferentes a la corteza frontal desde tegmento ponto-mesenceflico (locus coeruleus, rafe, sustancia negra) en la rata. An Anat 24:337–350

Reinoso-Surez F, Sierra G, Camacho A (1962) Efectos de lesiones en formación reticular protuberancial e istmo ponto-mesenceflico sobre el EEG del gato (Su participación en el mecanismo del sueño). Rev Med Univ Navarra 6:1–19
Reinoso-Surez F, Martínez-Moreno E, Nava BE (1974) Conexiones ascendentes desde el tronco del encéfalo a formaciones del hipocampo. An Anat 23:95–117
Reinoso-Surez F, Pasquier DA, Llamas A, Martínez-Moreno E (1975) Ascending connections to hypothalamus and prosencephalon from the pontine-mesencephalic tegmentum. Anat Rec 181:459–460
Reinoso-Surez F, Fairén A, Martínez-Moreno E, Nava BE, Pasquier DA (1977) Ascending projections from the brain stem. Med Biol Pap 1:233–246
Reinoso-Surez F, Llamas A, Avendaño C (1982) Pallido-cortical projections in the cat studied by means of the horseradish peroxidase retrograde transport technique. Neurosci Lett 29:225–229
Reinoso-Surez F, Rodrigo-Angulo ML, Rodríguez-Veiga E, De Andrés I (1990) Thalamic connections of oral pontine tegmentum sites whose cholinergic stimulation produces enhacement of paradoxical sleep signs. In: Mancia M, Marini G (eds) The diencephalon and sleep. Raven Press, New York, pp 49–63
Reinoso-Surez F, De Andrés I, Rodrigo-Angulo ML, Rodríguez-Veiga E (1994) Location and anatomical connections of a paradoxical sleep induction site in the cat ventral pontine tegmentum. Eur J Neurosci 6:1829–1836
Reinoso-Surez F, De Andrés I, Rodrigo-Angulo ML, Garzón M (2001) Brain structures and mechanisms involved in the generation of REM sleep. Sleep Med Rev 5:63–78
Richardson RT, DeLong MR (1990) Context-dependent responses of primate nucleus basalis neurons in a go/no-go task. J Neurosci 10:2528–2540
Rico B, Cavada C (1998) Adrenergic innervation of the monkey thalamus: an immunohistochemical study. Neuroscience 84:839–847
Rodrigo-Angulo ML, Reinoso-Surez F (1982) Topographical organization of the brainstem afferents to the lateral posterior-pulvinar thalamic complex in the cat. Neuroscience 7:1495–1508
Rodrigo-Angulo ML, Reinoso-Surez F (1988) Connections to the lateral posterior-pulvinar thalamic complex from the reticular and ventral lateral geniculate thalamic nuclei: a topographical study in the cat. Neuroscience 26:449–459
Rodrigo-Angulo ML, Reinoso-Surez F (1995) Afferent connections of the lateralis medialis thalamic nucleus in the cat. Brain Res Bull 38:53–67
Rodrigo-Angulo ML, Rodríguez-Veiga E, Reinoso-Surez F (2000) Serotonergic connections to the ventral oral pontine tegmentum. Implication in paradoxical sleep modulation. J Comp Neurol 418:93–105
Rodrigo-Angulo ML, Rodríguez-Veiga E, Reinoso-Surez F (2005) A quantitative study of the brainstem cholinergic projections to the ventral part of the oral pontine reticular nucleus (REM sleep induction site) in the cat. Exp Brain Res 160:334–343
Rodrigo-Angulo ML, Heredero S, Rodríguez-Veiga E, Reinoso-Surez F (2008) GABAergic and non-GABAergic thalamic, hypothalamic and basal forebrain projections to the ventral oral pontine reticular nucleus: their implication in REM sleep modulation. Brain Res 1210:116–125
Rossi GF, Zanchetti A (1957) The brain stem reticular formation. Anatomy and physiology. Arch Ital Biol 95:199–435
Rossi GF, Minobe K, Candia O (1963) An experimental study of the hypnogenic mechanisms of the brain stem. Arch Ital Biol 101:470–492
Rubio-Garrido P, Pérez-de-Manzo F, Porrero C, Galazo MJ, Clasc F (2009) Thalamic input to distal apical dendrites in neocortical layer 1 is massive and highly convergent. Cereb Cortex 19:2380–2395
Sakai K (1991) Physiological properties and afferent connections of the locus coeruleus and adjacent tegmental neurons involved in the generation of paradoxical sleep in the cat. Prog Brain Res 88:31–45

Sakai K, Sastre JP, Salvert D, Touret M, Tohyama M, Jouvet M (1979) Tegmentoreticular projections with special reference to the muscular atonia during paradoxical sleep in the cat: an HRP study. Brain Res 176:233–254
Sakurai T (2007) The neural circuit of orexin (hypocretin): maintaining sleep and wakefulness. Nat Rev Neurosci 8:171–181
Sakurai T, Amemiya A, Ishii M, Matsuzaki I, Chemelli RM, Tanaka H, Williams SC, Richardson JA, Kozlowski GP, Wilson S, Arch JR, Buckingham RE, Haynes AC, Carr SA, Annan RS, McNulty DE, Liu WS, Terrett JA, Elshourbagy NA, Bergsma DJ, Yanagisawa M (1998) Orexins and orexins receptors: a family of hypothalamic neuropeptides and G protein-coupled receptors that regulate feeding behavior. Cell 92:573–585
Sakurai T, Nagata R, Yamanaka A, Kawamura H, Tsujino N, Muraki Y, Kageyama H, Kunita S, Takahashi S, Goto K, Koyama Y, Shioda S, Yanagisawa M (2005) Input of orexin/hypocretin neurons revealed by a genetically encoded tracer in mice. Neuron 46:297–308
Sallmen T, Beckman AL, Stanton TL, Eriksson KS, Tarhanen J, Tuomisto L, Panula P (1999) Major changes in the brain histamine system of the ground squirrel Citellus lateralis during hibernation. J Neurosci 19:1824–1835
Sánchez-González MA, García-Cabezas MA, Rico B, Cavada C (2005) The primate thalamus is a key target for brain dopamine. J Neurosci 25:6076–6083
Saper CB, Scammell TE, Lu J (2005) Hypothalamic regulation of sleep and circadian rhythms. Nature 437:1257–1263
Sasaki K, Matsuda Y, Kawaguchi S, Mizuno N (1972) On the cerebello-thalamo-cerebral pathway for the parietal cortex. Exp Brain Res 16:89–103
Sasaki K, Kawaguchi S, Oka H, Sakai M, Mizuno N (1976) Electrophysiological studies on the cerebellocerebral projections in monkeys. Exp Brain Res 24:495–507
Sawaguchi T, Goldman-Rakic PS (1994) The role of D1-dopamine receptor in working memory: local injections of dopamine antagonists into the prefrontal cortex of rhesus monkeys performing an oculomotor delayed-response task. J Neurophysiol 71:515–528
Schiff ND (2008) Central thalamic contributions to arousal regulation and neurological disorders of consciousness. Ann NY Acad Sci 1129:105–118
Schwartz JC, Arrang JM, Garbarg M, Pollard H, Ruat M (1991) Histaminergic transmission in the mammalian brain. Physiol Rev 71:1–51
Semba K (1993) Aminergic and cholinergic afferents to REM sleep induction regions of the pontine reticular formation in the rat. J Comp Neurol 330:543–556
Semba K (1999) The mesopontine cholinergic system: a dual role in REM sleep and wakefulness. In: Lydic R, Baghdoyan HA (eds) Handbook of behavioural state control. Cellular and molecular mechanisms. CRC Press, Boca Raton, London, New York, pp 161–180
Semba K (2000) Multiple output pathways of the basal forebrain: organization, chemical heterogeneity, and roles in vigilance. Behav Brain Res 115:117–141
Sforza E, Montagna P, Tinuper P, Cortelli P, Avoni P, Ferrillo F, Petersen R, Gambetti P, Lugaresi E (1995) Sleep-wake cycle abnormalities in fatal familial insomnia. Evidence of the role of the thalamus in sleep regulation. Electroencephalogr Clin Neurophysiol 94:398–405
Sherin JE, Elmquist JK, Torrealba F, Saper CB (1998) Innervation of histaminergic tuberomammillary neurons by GABAergic and galaninergic neurons in theventrolateral preoptic nucleus of the rat. J Neurosci 18:4705–4721
Sherman SM (2001) Tonic and burst firing: dual modes of thalamocortical relay. Trends Neurosci 24:122–136
Sherman SM (2004) Interneurons and triadic circuitry of the thalamus. Trends Neurosci 27:670–675
Shirvalkar P, Seth M, Schiff ND, Herrera DG (2006) Cognitive enhancement with central thalamic electrical stimulation. Proc Natl Acad Sci USA 103:17007–17012
Siegel JM, Tomaszewski KS, Nienhuis R (1986) Behavioral states in the chronic medullary and midpontine cat. Electroencephalogr Clin Neurophysiol 63:274–288

Sillito AM, Jones HE, Gerstein GL, West DC (1994) Feature-linked synchronization of thalamic relay cell firing induced by feedback from the visual cortex. Nature 369:479–482

Sofroniew MV, Priestley JV, Consolazione A, Eckenstein F, Cuello AC (1985) Cholinergic projections from the midbrain and pons to the thalamus in the rat, identified by combined retrograde tracing and choline acetyltransferase immunohistochemistry. Brain Res 329:213–223

Spruston N (2008) Pyramidal neurons: dendritic structure and synaptic integration. Nat Rev Neurosci 9:206–221

Starzl TE, Magoun HW (1951) Organization of the diffuse thalamic projection system. J Neurophysiol 14:133–146

Steininger TL, Alam MN, Gong H, Szymusiak R, McGinty D (1999) Sleep-waking discharge of neurons in the posterior lateral hypothalamus of the albino rat. Brain Res 840:138–147

Steriade M (1981) Mechanisms underliyng cortical activation: neuronal organization and propierties of the midbrain reticular core and intralaminar thalamic nuclei. In: Pompeiano O, Ajmone-Marsan C (eds) Brain mechanisms of perceptual awareness and purposeful behavior, vol 8, IBRO monogragh series. Raven Press, New York, pp 327–377

Steriade M (2004) Acetylcholine systems and rhythmic activities during the waking-sleep cycle. Prog Brain Res 145:179–196

Steriade M, Domigh L, Oakson G, Deschenes M (1987a) The deafferented reticular thalamic nucleus generates spindle ritmicity. J Neurophysiol 57:260–273

Steriade M, Parent A, Pare D, Smith Y (1987b) Cholinergic and non-cholinergic neurons of cat basal forebrain project to reticular and mediodorsal thalamic nuclei. Brain Res 408:372–376

Steriade M, Paré D, Parent A, Smith Y (1988) Projections of cholinergic and non-cholinergic neurons of the brainstem core to relay and associational thalamic nuclei in the cat and macaque monkey. Neuroscience 25:47–67

Steriade M, Paré D, Datta S, Oakson G, Curró Dossi R (1990) Different cellular types in mesopontine cholinergic nuclei related to ponto-geniculo-occipital waves. J Neurosci 10:2560–2579

Steriade M, Curró Dossi R, Núñez A (1991) Network modulation of a slow intrinsic oscillation of cat thalamocortical neurons implicated in sleep delta waves: cortically induced synchronization and brainstem cholinergic suppression. J Neurosci 11:3200–3217

Steriade M, Curro-Dossi R, Contreras D (1993) Electrophysiological properties of intralaminar thalamocortical cells discharging rhythmic (approximately 40 HZ) spike-bursts at approximately 1000 HZ during waking and rapid eye movement sleep. Neuroscience 56:1–9

Sterman MB, Clemente CD (1962) Forebrain inhibitory mechanisms: cortical synchronization induced by basal forebrain stimulation in the behaving cat. Exp Neurol 6:91–102

Sutcliffe JG, de Lecea L (2002) The hypocretins: setting the arousal threshold. Nat Rev Neurosci 3:339–349

Sutin EL, Jacobowitz DM (1988) Immunocytochemical localization of peptides and other neurochemicals in the rat laterodorsal tegmental nucleus and adjacent area. J Comp Neurol 270:243–270

Szymusiak R, McGinty D (1989) Effects of basal forebrain stimulation on the waking discharge of neurons in the midbrain reticular formation of cats. Brain Res 498:355–359

Takahashi K, Lin JS, Sakai K (2006) Neuronal activity of histaminergic tuberomammillary neurons during wake-sleep states in the mouse. J Neurosci 26:10292–10298

Tao R, Ma Z, McKenna JT, Thakkar MM, Winston S, Strecker RE, McCarley RW (2006) Differential effect of orexins (hypocretins) on serotonin release in the dorsal and median raphe nuclei of freely behaving rats. Neuroscience 141:1101–1105

Tashiro M, Sakurada Y, Mochizuki H, Horikawa E, Maruyama M, Okamura N, Watanuki S, Arai H, Itoh M, Yanai K (2008) Effects of a sedative antihistamine, D-chlorpheniramine, on regional cerebral perfusion and performance during simulated car driving. Hum Psychopharmacol 23:139–150

Thakkar MM, Strecker RE, McCarley RW (1998) Behavioral state control through differential serotonergic inhibition in the mesopontine cholinergic nuclei: a simultaneous unit recording and microdialysis study. J Neurosci 18:5490–5497

Thakkar MM, Ramesh V, Strecker RE, McCarley RW (2001) Microdialysis perfusion of orexin-A in the basal forebrain increases wakefulness in freely behaving rats. Arch Ital Biol 139:313–328

Tortelly A, Reinoso-Surez F (1980) Projections to the superior colliculus from the dorsal hypothalamic area and other prosencephalic structures derived from the embryonic subthalamic longitudinal band of the diencephalon. Neurosci Lett 18:257–260

Torterolo P, Sampogna S, Morales FR, Chase MH (2006) MCH-containing neurons in the hypothalamus of the cat: searching for a role in the control of sleep and wakefulness. Brain Res 1119:101–114

Torvik A, Brodal A (1957) The origin of reticulospinal fibers in the cat; an experimental study. Anat Rec 128:113–137

Tsujino N, Yamanaka A, Ichiki K, Muraki Y, Kilduff TS, Yagami K, Takahashi S, Goto K, Sakurai T (2005) Cholecystokinin activates orexin/hypocretin neurons through the cholecystokinin A receptor. J Neurosci 25:7459–7469

Ursin R, Sterman MB (1981) A manual for standardized scoring of sleep and waking states in the adult cat. University of California Brain Information Service, Los Angeles

van den Pol AN, Acuna-Goycolea C, Clark KR, Ghosh PK (2004) Physiological properties of hypothalamic MCH neurons identified with selective expression of reporter gene after recombinant virus infection. Neuron 42:635–652

Vanni-Mercier G, Sakai K, Jouvet M (1984) Specific neurons for wakefulness in the posterior hypothalamus in the cat. CR Acad Sci III 298:195–200

Vanni-Mercier G, Sakai K, Lin JS, Jouvet M (1991) Carbachol microinjections in the mediodorsal pontine tegmentum are unable to induce paradoxical sleep after caudal pontine and prebulbar transections in the cat. Neurosci Lett 130:41–45

Velasco M, Lindsley DB (1965) Role of orbital cortex in regulation of thalamocortical electrical activity. Science 149:1375–1377

Velayos JL (1971) Conexiones ascendentes del bulbo raquídeo. An Anat 20:587–627

Velayos JL, Reinoso-Surez F (1982) Topographic organization of the brainstem afferents to the medio-dorsal thalamic nucleus. J Comp Neurol 206:12–28

Velayos JL, Reinoso-Surez F (1985) Prosencephalic afferents to the mediodorsal thalamic nucleus. J Comp Neurol 242:161–181

Velayos JL, Jiménez-Castellanos J Jr, Reinoso-Surez F (1989) Topographical organization of the projections from the reticular thalamic nucleus to the intralaminar and medial thalamic nuclei in the cat. J Comp Neurol 279:457–469

Velayos JL, Casas-Puig R, Reinoso-Surez F (1993) Laminar organization of the cortical projections to the intralaminar and medial thalamic nuclei. In: Minciacchi D, Molinari M, Macchi G, Jones EG (eds) Thalamic networks for relay and modulation. Pergamon Press, Oxford, pp 185–195

Verret L, Goutagny R, Fort P, Cagnon L, Salvert D, Léger L, Boissard R, Salin P, Peyron C, Luppi PH (2003) A role of melanin-concentrating hormone producing neurons in the central regulation of paradoxical sleep. BMC Neurosci 4:19

Villablanca JR (1965) The electrocorticogram in the chronic *cerveau isolé* cat. Electroenceph clin Neurophysiol 19:576–586

Villablanca JR (1972) Permanent reduction in sleep after removal of cerebral cortex and striatum in cats. Brain Res 36:463–468

Villablanca JR (1974) Role of the thalamus in sleep control: sleep-wakefulness studies in chronic diencephalic and athalamic cats. In: Petre-Quadens O, Schlag J (eds) Basic sleep mechanisms. Academic, New York, pp 51–81

Villablanca JR (2004) Counterpointing the functional role of the forebrain and of the brainstem in the control of the sleep-waking system. J Sleep Res 13:179–208

Villablanca JR, Marcus R (1972) Sleep-wakefulness, EEG and behavioral studies of chronic cats without neocortex and striatum: the 'diencephalic' cat. Arch Ital Biol 110:348–382
Vincent SR, Hope BT (1992) Neurons that say NO. Trends Neurosci 15:108–113
Viñes-Morros R (1959) Vias trigeminales y actividad bioeléctrica cerebral. An Anat 10:507–527
von Economo C (1926) Die Pathologie des Schlafes. In: von Bethe A, von Bergman G, Embden G, Ellinger A (eds) Handbuch des normalen und pathologischen Physiologie. Springer, Berlin, pp 591–610
von Economo C (1930) Sleep as a problem of localization. J Nerv Ment Dis 71:249–259
Wada H (1984) Distribution of the histaminergic neuron system in the central nervous system of rats; a fluorescent immunohistochemical analysis with histidinedecarboxylase as a marker. Brain Res 295:13–25
Wada H, Inagaki N, Yamatodani A, Watanabe T (1991) Is the histaminergic neuron system a regulatory center for whole-brain activity? Trends Neurosci 14:415–418
Wang QP, Koyama Y, Guan JL, Takahashi K, Kayama Y, Shioda S (2005) The orexinergic synaptic innervation of serotonin- and orexin 1-receptor-containing neurons in the dorsal raphe nucleus. Regul Pept 126:35–42
Watanabe T, Taguchi Y, Shiosaka S, Tanaka J, Kubota H, Terano Y, Tohyama M, Wada H (1984) Distribution of the histaminergic neuron system in the central nervous system of rats; a fluorescent immunohistochemical analysis with histidine decarboxylase as a marker. Brain Res 295:13–25
Westlund KN, Sorkin LS, Ferrington DG, Carlton SM, Willcockson HH, Willis WD (1990) Serotoninergic and noradrenergic projections to the ventral posterolateral nucleus of the monkey thalamus. J Comp Neurol 295:197–207
Williams SM, Goldman-Rakic PS (1993) Characterization of the dopaminergic innervation of the primate frontal cortex using a dopamine-specific antibody. Cereb Cortex 3:199–222
Williams JA, Reiner PB (1993) Noradrenaline hyperpolarizes identified rat mesopontine cholinergic neurons in vitro. J Neurosci 13:3878–3883
Wilson FAW, Rolls ET (1990) Neuronal responses related to novelty and familiarity of visual stimuli in the substantia innominata, diagonal band of Broca and periventricular region of the primate basal forebrain. Exp Brain Res 80:104–120
Wilson JR, Manning KA, Forestner DM, Counts SE, Uhlrich DJ (1999) Comparison of cholinergic and histaminergic axons in the lateral geniculate complex of the macaque monkey. Anat Rec 255:295–305
Xi MC, Morales FR, Chase MH (2001) Effects on sleep and wakefulness of the injection of hypocretin-1 (orexin-A) into the laterodorsal tegmental nucleus of the cat. Brain Res 901:259–264
Yamanaka A, Tsujino N, Funahashi H, Honda K, Guan JL, Wang QP, Tominaga M, Goto K, Shioda S, Sakurai T (2002) Orexins activate histaminergic neurons via the orexin 2 receptor. Biochem Biophys Res Commun 290:1237–1245
Yamanaka A, Muraki Y, Ichiki K, Tsujino N, Kilduff TS, Goto K, Sakurai T (2006) Orexin neurons are directly and indirectly regulated by catecholamines in a complex manner. J Neurophysiol 96:284–298
Yoshida K, McCormack S, España RA, Crocker A, Scammell TE (2006) Afferents to the orexin neurons of the rat brain. J Comp Neurol 494:845–861
Zanchetti A (1967) Brain stem mechanisms of sleep. Anestesiology 28:81–99
Zarranz J (1972) Nuevas experiencias sobre la región caudal del puente y su influencia en los mecanismos neurofisiológicos del sueño. An Anat 21:583–611
Zarranz J, Reinoso-Surez F (1971) The pontine tegmentum and sleep-wakefulness states. XXV International Congress Physiology Science, Munich, p 618
Zeitzer JM, Buckmaster CL, Parker KJ, Hauck CM, Lyons DM, Mignot E (2003) Circadian and homeostatic regulation of hypocretin in a primate model: implications for the consolidation of wakefulness. J Neurosci 23:3555–3560
Zernicki B (1968) Pretrigeminal cat. Brain Res 9:1–15

Subject Index

MIX
Papier aus verantwortungsvollen Quellen
Paper from responsible sources
FSC® C105338

If you have any concerns about our products,
you can contact us on
ProductSafety@springernature.com

In case Publisher is established outside the EU,
the EU authorized representative is:
Springer Nature Customer Service Center GmbH
Europaplatz 3, 69115 Heidelberg, Germany

Printed by Libri Plureos GmbH
in Hamburg, Germany